KT-439-339

MARIE CURIE

and Her

DAUGHTERS

THE PRIVATE LIVES
OF SCIENCE'S
FIRST FAMILY

SHELLEY EMLING

palgrave
macmillan

MARIE CURIE AND HER DAUGHTERS
Copyright © Shelley Emling, 2012
All rights reserved.

First published in hardcover in 2012 by PALGRAVE MACMILLAN® in the
US—a division of St. Martin's Press LLC, 175 Fifth Avenue, New York, NY
10010.

Where this book is distributed in the UK, Europe and the rest of the world, this
is by Palgrave Macmillan, a division of Macmillan Publishers Limited, registered
in England, company number 785998, of Houndmills, Basingstoke, Hampshire
RG21 6XS.

Palgrave Macmillan is the global academic imprint of the above companies and
has companies and representatives throughout the world.

Palgrave® and Macmillan® are registered trademarks in the United States, the
United Kingdom, Europe and other countries.

ISBN: 978-1-137-27836-4

Library of Congress Cataloging-in-Publication Data is available from the Library
of Congress.

A catalogue record of the book is available from the British Library.

Design by Greg Collins

First PALGRAVE MACMILLAN paperback edition: August 2013

10 9 8 7 6 5 4 3 2 1

Printed in the United States of America.

*This book is dedicated with love to my son Chris,
who is so much more than I was at his age.*

*It is also dedicated to girls and women everywhere
who are studying—or are already succeeding
in—the fields of math and science.*

Praise for *Marie Curie and Her Daughters*

"A must-read for every woman and every female teenager. In accessible prose, Emling enlightens the world about this enigmatic scientist, and, with the help of personal letters shared by Curie's granddaughter, Emling has woven a story of a woman full of grace and of the daughters who loved her without fail. I loved this book."

—*Mary H. Manhein, author of* The Bone Lady *and* Trail of Bones

"Shelley Emling's dazzling chronicle of the three Curies and their world-famous accomplishments is surpassed only by her account of how each stretched her era's concept of the possibilities for women. A tour de force!"

—*Megan McKinney, author of* The Magnificent Medills

"Ms. Emling's riveting new biography reveals in page-turning prose the life-balance struggles of a true genius. It's a tip of the hat to the private Marie, the single working mother, whose many accomplishments include her two amazing daughters."

—*Lisa Verge Higgins,* New York Journal of Books

"Shelley Emling's excellent joint biography of Marie, her daughters (and a granddaughter, too) is an exhilarating story that couples scientific discovery and motherhood. A book that should propel young women into science for the sheer fun of it, it's also a rich tale of war and peace, family commitment, friendship, and medical progress."

—*Adele Glimm, author of* Gene Hunter *and* Rachel Carson

30131 05325838 7

LONDON BOROUGH OF BARNET

MARIE CURIE

and Her

DAUGHTERS

CONTENTS

Eight pages of photographs appear between pages 104 and 105.

ACKNOWLEDGMENTS

The first time I ever heard from Hélène Langevin-Joliot was via email on January 3, 2011. Opening up my inbox, I did a double take as I glanced down and noticed the name of the granddaughter of one of the most famous women who ever lived. She was writing to say that she'd be available to meet with me the week of April 18. I was over the moon. When I did finally meet with her at the Marie Curie Institute in Paris that week, I was as starstruck as if I were meeting the president. She was as kind and helpful and gracious as she possibly could have been. She made sure I knew that her grandmother never sought to succeed in a male-dominated arena; rather, she simply loved science above everything else. I want to thank her for sharing her insights about her family and also about America's impact on her grandmother. She referred me to a selection of more than two hundred letters, in French, exchanged between Marie Curie and her daughters as well as to other papers and documents including eighty-eight pages of Irene Curie's own remembrances, also in French. Langevin-Joliot has had those letters published in France in a book titled *Marie-Curie et ses filles: Lettres*.

I also want to thank Renaud Huynh, director of the Curie Institute, for offering his own insights and for answering all my pesky questions. And thanks to Jocelyn Wilk and other employees for their assistance at the Columbia University Archives.

Personally I'd like to thank my agent, the indefatigable Agnes Birnbaum. Thank you to my incredible editor at Palgrave Macmillan, Luba Ostashevsky, who patiently helped me better grasp the art of narrative history. Also thanks to Laura Lancaster, Victoria Wallis, and Georgia Maas, and others at Palgrave for all their hard work on my behalf.

There is no way I could have completed this book without some extremely talented French translators. Thank you to Florence Sinofsky, Louis Gagnon, Isabelle de Carville, and, especially Kim Parkash. Kim, in particular, worked tirelessly to translate letter after letter for me. If there are any errors in this book, they are mine and mine alone.

Thank you also to my bosses at AOL Patch for granting me a leave of absence.

Over the years, the Curies have garnered the attention of countless writers who have meticulously researched the lives of the various family members. Without them, it would be impossible for someone like me to tell my own story of the Curie women in my own way. They include Denis Brian, Barbara Goldsmith, Sarah Dry, Robert Reid, Naomi E. Pasachoff, Susan Quinn, Rosalynd Pflaum, Michel Pinault, Maurice Goldsmith and, of course, Eve Curie, who wrote a wonderful biography of her mother and also an astonishing book about her adventures as a World War II correspondent. Alan E. Waltar wrote a fascinating book on radiation and modern life. And Marie Curie wrote a biography of her husband that included a short autobiography.

I am indebted to my husband, Scott, for reading every chapter. His intelligence never ceases to amaze me. My appreciation also goes to my three gorgeous children, Chris, Ben, and Olivia, who always showed an interest in the book's progress. I promised them I would thank Pepper—my constant companion—as well. Thanks to my girlfriends who bring me such joy. And a special thanks to my mother, Lois Ruth. She was, every day of her life, my greatest cheerleader.

Finally, I drew so much inspiration from five women—Marie Curie, Irene Curie, Eve Curie, Missy Meloney, and Hélène Langevin-Joliot—who had and have many wonderful qualities. But the quality I love most is that none of them ever—ever—waited for something to happen to them. They made things happen for themselves. My hope is that girls and women everywhere will do the same.

America

A Fresh Point of Departure for the
World's Greatest Scientist

The faint outline of New York City emerged through a haze of warm weather on the morning of May 11, 1921, as the RMS *Olympic* steamed its way into New York Harbor, the last leg of its weeks-long journey across the Atlantic. On board was the world's most famous scientist, Madame Marie Curie, accompanied by an unlikely companion, a feisty American journalist named Missy Meloney. Meloney had warned Marie that a mob of reporters and photographers would be gathered on the landing pier, anxiously awaiting the arrival of the one they dubbed the "benefactress of the human race." Marie, frail and mortified by publicity, took her time making an appearance before a crowd that had already been waiting a good five hours. Finally, Meloney managed to set her up comfortably in a giant armchair—like a queen on a throne—on the boat deck of the ship.

From a menacingly close distance, hordes of photographers began snapping away with abandon. "Look this way, Madame Curie! Turn your head to the right! Lift your head! Look this way! This way! This way!" The incessant clicking of machines created a cacophony of chaos that assaulted the senses of the astonished fifty-three-year-old physicist traveling outside Europe for the first time in her life. Looking on and acting as bodyguards were Marie's two daughters, Irene and Eve, aged twenty-three and sixteen at the time. The three Curie women traveled with only one trunk of clothing.

Following a hasty press conference, Marie no doubt was relieved to finally enjoy a reprieve from the noise at Meloney's New York City apartment, which she and her daughters used as a base during a busy lineup of speeches, luncheons, and ceremonies over the next seven weeks. Everywhere they went, another honor, medal, degree, or superlative was bestowed upon her. As Eve Curie later wrote, "Americans had surrounded Mme Curie with an almost religious devotion and had placed her in the first rank of living men and women." Each day was more stimulating than the one before. Innumerable strangers stared at Marie, with everyone jostling each other to get a better view. At one stage, a fan shook Marie's hand so enthusiastically that she later had to have her arm bandaged. Simply put, America couldn't get enough of her.

But it would all be worth it.

Nine days after arriving in America, at exactly 4 P.M. on May 20, 1921, the double doors to the East Room, the largest room in the White House, opened slowly, signaling the start of a grand entrance. Leading the way was First Lady Florence Harding alongside Jules Jusserand, the French ambassador, followed by Madame Curie on the arm of President Warren Harding. Finally, there was Missy Meloney, next to Irene and Eve Curie and trailed by the many American ladies of the Marie Curie Radium Committee.

Marie was about to get what she had come for—a single gram of the precious radium she had discovered in 1898, purchased with the enthusiastic help of women and young girls from all across America. Less obvious was that the plainly dressed woman, who had never been fully appreciated in Europe, also was about to get a new sense of self, one that would carry her into the final chapter of her life. So momentous was Marie's trip to America that its success continued in one incarnation or another for the remainder of her days on earth—with many, many highlights to come. As Eve Curie put it, a veil fell away during the journey, allowing her and her sister to see what their "sweet Mé" actually meant to the world.

In the United States, the post–World War I recession was easing up and the country was entering a time of prosperity. The newspapers were touting Harding's motto: "Less government in business and more business in government." Women were feeling powerful after gaining the right to vote in 1920.

Whether they liked it or not, America's vitality was already rubbing off on the Curie family. Marie was seen as the embodiment of the people's own hopes and dreams—a brilliant career woman and doting mother who was being celebrated, at long last, in the most elaborate way possible.

Like so many others, I learned about Madame Curie when I was a young student—about how she had risen from a modest Polish family to study at the Sorbonne. And about how she had become the first person to be honored with two Nobel Prizes, an achievement especially extraordinary for a woman of her time. I'm sure someone must have told me she discovered not one but two new elements—both radium and polonium—and that she was the first to use the word "radioactivity." Although known primarily for her discovery of radium, her true gift to science, according to biographer Roger M. Macklis, was the realization that this radioactivity is an intrinsic atomic property of matter rather than the result of more superficial chemical processes. Under her personal direction, the world's first experiments in treating neoplasms—or tumors—with radioactive isotopes were completed. With the possible exception of Albert Einstein, she remains, arguably, the most famous scientist in history. Her life reflected loyalty, generosity, dogged determination, an unwavering focus on work, and a strong belief in the benefits of pure science—all traits especially admired in America. Most amazing to me is that, as a young woman, she had faith that there was something to be found even when she didn't know what she was looking for. After hypothesizing about the existence of radioactive substances, she spent years seeking out these substances by sifting through—literally—tons of a discarded uranium ore called pitchblende, working in a leaky wooden shed under hazardous conditions. Indeed, according to writer Denise Ham, Marie—with her fingertips burned and cracked—was able to extract only four decigrams of radium chloride (about the weight of four postage stamps) from one ton of pitchblende.

Marie is also a tragic figure, having been widowed at the age of thirty-eight when her scientist husband, Pierre Curie, died after being run over by a horse-drawn wagon in Paris in 1906.

But little has been taught or written about Marie's friendship with the American journalist Mrs. William B. Meloney, nicknamed Missy, who was the editor of a famous magazine called *The Delineator*. That friendship turned around the lives of both Marie and Meloney and lead to a nationwide fundraiser by American women to buy radium for Marie's research. Meloney was everything Marie was not—a fireball and known talker who was packed with charisma. And she had a persistence to match Marie's own. Woe to anyone who stood in her way. Meloney not only was close to Marie; she also adored Marie's two daughters, who were both notoriously protective of their famous mother.

It was thanks to Meloney's tutelage that Marie was, to a large extent, able to overcome her aversion to publicity in order to gain access to the financial and scientific resources of the United States where, as Eve Curie put it, "Nothing is impossible." At the time, the scientific centers in the United States together had about fifty times as much radium as the single gram she—the scientist who had discovered the element—safeguarded in her inadequate laboratory in France.

During her journey through America, orchestrated by Meloney, Marie was able to realize for the first time how her celebrity status could empower her to have an impact on the causes she favored. In the years to come, she would speak before large crowds at meetings and conferences throughout the world, gradually becoming more comfortable in the spotlight. Marie discovered that people were very willing to support her work, and she went on to enjoy great success as a fundraiser for the Radium Institute she had founded in Paris in 1914 and then for her Radium Institute in Warsaw. Later, she also lent her name to the cause for world peace by working for the League of Nations. It was in large part thanks to Meloney's influence. Before meeting the journalist, Marie had abhorred any kind of publicity. "[Curie], who handles daily a particle of radium more dangerous than lightning, was afraid when confronted by the necessity of appearing before the public," a French newspaper editor once remarked.

Following many lengthy periods apart, Marie's trek through America also allowed her some real quality time with her girls. They would both turn out to be extraordinary women, with Irene going on to win her own Nobel Prize and Eve becoming a celebrated author and war correspondent. Eve, a French national, became an American citizen in 1958 after marrying the American politician and diplomat Henry Richardson Labouisse Jr. in November 1954. She spent her last years in New York City, where she died at the age of 102 in 2007.

Many books on Madame Curie tell the stories of her humble childhood and her extraordinary collaboration with her husband. Indeed, her years as a teenager working as a governess while staying up nights to study math and science are the stuff of legend. However, even the weightiest tomes often tidily wrap up the story of her life around 1911, when Marie received her second Nobel Prize. But America's love affair with Marie Curie—and Marie's trips to the United States (she was to return in 1929)—were momentous occasions, not only for Marie and her daughters, but also for the American people. Marie had found a patron while America had found a woman role model and, equally important, a cause to rally around. The hospitality and generosity showered on the three Curie women, whichever city they found themselves in as they crisscrossed the country from Boston to Flagstaff, were evidence of just how passionate America was about this foreigner and her family. As soon as Meloney spread the word that Marie needed radium, countless Americans—from prominent doctors to modest students—got behind a fund-raising campaign in a tangible manner. The movement was not unlike the similarly monumental fund-raising drive that had unfolded in the late 1800s when America sought to build a base for the Statue of Liberty, a gift presented by the people of France. At that time, too, much of the country was swept up in raising money. Marie's second American tour in 1929 also proved bountiful, successfully equipping the Warsaw Radium Institute founded by Marie in 1925 with her sister Bronya as director.

There seemed to be so much more to learn about Madame Curie's life. I ventured to write another biography about a person who's been written about so many times before in order to look at the woman, mother, and friend behind the pioneering scientist. Much of the research for this biography was made possible through the generosity of Marie's only granddaughter, Hélène Langevin-Joliot, and the Marie Curie Institute in Paris. The octogenarian nuclear physicist gave me eighty-eight pages of remembrances written by her mother Irene about her grandmother. She also led me to a collection of more than two hundred letters in French exchanged between Marie and her two daughters from 1906 to 1934. A few years ago, Langevin-Joliot also made public many boxes of archives relating to both her mother and her father. Combined, this new information opens a window into what Marie was like as a mother and how she walked a tightrope between family and career. Marie's absorption with work meant an inordinate amount of time spent away from home but it never prevented her from being intimately involved in the development of her daughters. At one point, Eve Curie wrote, Marie was so depressed she was suicidal. But she never lost sight of her goals for her children.

Letters show that the three often swapped the most trivial of details. In one, the health of a goldfish is discussed; in another, the height of waves at the beach. Yet they also reveal a mother who was always apologizing for missing her children's birthdays but who was so dead set on developing their minds that she constantly sent them complicated math problems to solve.

Langevin-Joliot's own insights were also extremely helpful.

"My grandmother wasn't the same person when she came back from America . . . the trip changed her forever," the distinguished woman told me. Sitting among stacks of papers in a small office at the Marie Curie Institute, she explained how most books covered her grandmother's life from childhood to World War I as though "she suddenly died after that and nothing else ever happened." She made a point of emphasizing how, at first, her grandmother had been known only as the wife of a great scientist and then later only as the humble Polish girl who endured great trials to become a renowned scientist. "But there was a lot more to her than that . . . she was a mother, a peacemaker, a humanitarian," she said.

The other person who confirmed the validity of my mission was Renaud Huynh, the director of the Curie Institute in Paris, who agreed that the last twenty years of Curie's life were extremely important—but have generally been overlooked. "Writers often forget that the life of Marie Curie does not stop at the end of World War I and that the Radium Institute she helped create quickly became a place of international reputation which continues to this day," he said. "The end of her life—from 1918 to 1934—is largely unknown to readers." Langevin-Joliot nodded in agreement, adding that "I think the idea of writing about the last twenty years of her life and beyond is something that is never done."

And, with those words, let's head back to that day in May 1921 when the East Room of the White House was teeming with more than one hundred of the world's most distinguished scientists and diplomats from Poland, France, and the United States. Overseeing them all was the affable President Harding who, having suffered multiple scandals during his administration, was relishing the distraction of the pleasant afternoon ceremony.

To enthusiastic applause, the president presented Marie with a golden key inscribed "From the Women of America to Madame Marie Curie." Later Marie would open a lead-lined casket containing her single gram of radium. Harding paid homage to Marie, to her adopted nation, France, and to the newly re-created nation and land of her birth, Poland, which had finally become an independent country again after World War I:

"As a nation whose womanhood has been exalted to fullest participation in citizenship, we are proud to honor in you a woman whose work has earned universal acclaim and attested woman's equality in every intellectual and spiritual activity."

After his speech, Marie stood to respond:

"I cannot express to you the emotion which fills my heart in this moment. You, the chief of this great Republic of the United States, honor me as no woman has ever been honored in America before. I accept this rare gift, Mr. President, with the hope that I may make it serve mankind. I thank your countrywomen in the name of France. I thank them in the name of humanity, which we all wish so much to make happier. I love you all, my American friends, very much."

Realizing that Madame Curie was in frail health, Mrs. Harding guided the scientist to a chair where she could rest before sending the gathering of dignitaries over to Irene and Eve. The two mingled with the guests, comfortably conversing with them in English, French, or Polish, or a mixture of all three. The young ladies had been carefully raised with the confidence to speak on their mother's behalf when necessary.

All told, this brief time at the White House marked the high point of Marie's triumphal but physically and emotionally draining journey through America. It was particularly noteworthy that Marie even chose to end her autobiography—written at Meloney's urging—with a description of the 1921 trip. In Marie's own words:

"I got back to France with a feeling of gratitude for the precious gift of the American women, and with a feeling of affection for their great country tied with ours by mutual sympathy which gives confidence for a peaceful future for humanity."

And so it was a new beginning for Marie, just when many believed her life had reached its culmination. And, most importantly, the 1921 trip marked a dramatic comeback for a hero who, unbeknownst to many today, had hit rock bottom just ten years before.

CHAPTER 1

An Absolutely Miserable Year

M adame Marie Skłodowska Curie had come to Brussels in the first few days of November 1911 to talk physics with her peers—and also to escape. She was the only woman among twenty-three men attending a gathering of some of the world's greatest minds that included Albert Einstein and Max Planck. When they weren't debating the challenge to modern physics presented by the discovery of radioactivity, the scientists surely talked among themselves about the rumors of illicit romance swirling around the forty-three-year-old widow. In one photograph of the delegation, Marie can be seen sitting at a table in the front, with her head down over some paperwork, while her former professor, the great mathematician and physicist Henri Poincaré, looks on. Behind them stands her married lover, Paul Langevin, thirty-eight, a dashing father of four and an expert on molecular and kinetic theory.

The wealthy industrialist and philanthropist Ernest Solvay had organized the all-expenses-paid week-long conference to focus on "the theory of radiation and quanta." He had invited Marie and the others in June—just as her life was starting to unravel. Indeed, his invitation couldn't have come at a better time. After all she had discovered, after all she had achieved, Marie's presence was no longer welcomed in France even by some who had once been among her greatest admirers. Photos resembling mug shots appeared in newspapers in an anti-Semitic broadside.

Adversaries showed up at her home, hurling rocks at the windows. Fellow professors at the Sorbonne wanted the university's first female teacher out. The mother of two was close to a nervous breakdown, on the brink of madness. The Solvay Conference was a welcome diversion.

Almost all of those invited had played a prominent part in the advancement of major scientific theories. And perhaps no one more so than Marie, who had discovered two important elements, polonium and radium. A few, such as the prominent Sorbonne professor Jean Perrin, Marie had known quite well and for many years. Others, such as Einstein, she was meeting for the first time.

Complaining of nagging headaches, she walked out of several committee meetings before they were over. But her obvious distress didn't stop her from discussing with other scientists the creation of an international standard—called the Curie, in memory of Pierre Curie—that could be used to compare radium preparations from different countries. Preparation of the radium standard, which is still in use today, had been assigned to Marie, who argued in the face of some opposition that a Curie must correspond to more than just an infinitesimal amount of radium.

Worn down with worry, Marie was handed a telegram midway through the conference. The modest woman who hated attention was almost afraid to open it. But the news was not what she expected: "Nobel Prize for chemistry awarded to you. Letter follows." It could just as easily have been talking about the weather for all its simplicity. But this message from Carl Aurivillius, head of the Swedish Committee on Prizes, confirmed her place in history. Marie was about to become the first person—man or woman—to be awarded two Nobel Prizes. To this day she remains the only person to have been awarded two Nobel science prizes in different subjects. (Linus Pauling is the only person to have been awarded two unshared Nobel Prizes—the 1954 Nobel Prize in Chemistry and the 1962 Nobel Peace Prize.)

As news of the second Nobel Prize circulated, some at the meeting grumbled to themselves that, in essence, Marie had been awarded the same prize twice, since both were related to her work on radioactivity. But praise for her years of research also flowed forth, and many of the gentlemen in attendance were gracious enough to offer their heartfelt congratulations. And yet the realization remained that despite being awarded a second Nobel Prize—an unprecedented feat for anyone but especially for a woman in a field that remained the bastion of men—1911 was a year of humiliation, depression, and defeat for Marie. Despite so many accomplishments, her star had been falling fast during the last several months and she'd become a woman on the edge, close to losing it all.

Her life's downward spiral had begun almost a year before when, in December 1910, she had decided to emulate her late husband, Pierre, by competing for a single open seat in the French Academy of Sciences, an institution that held sway over the support and direction of French science. Although the prestigious organization boasted only 68 members, the idea of a woman trying to break into the male stronghold sparked so much attention that all 163 members of the French Institute—the umbrella organization representing five different academies—showed up to have their say. Despite the ringing endorsement of the respected newspaper *Le Figaro* which had named her the nation's most famous physicist the election on January 24, 1911 did not go as Marie had hoped. There were all sorts of reasons for this, but not one of them made sense. Driven by the country's growing xenophobia, some members circulated a bizarre charge that Marie's application for membership had actually been contrived by a Jewish cabal to block the honor from going to an equally talented but more Catholic candidate. Next, a conservative periodical fanned the flames by claiming that, with her Polish heritage and a maiden name like Skłodowska, Marie almost certainly was a Jew herself. (She wasn't: her mother was Catholic and her father an atheist.) Another publication splashed two photos of her on its cover that looked like police mug shots, making it seem as though she were a criminal on the run. In the weeks leading up to the vote, an all-out smear campaign took on a life of its own while willfully ignoring Marie's impeccable résumé. No doubt sexism was a major culprit. In the end, Marie lost by only two votes to radio pioneer Édouard Branly, an inventor who had been honored by the pope and backed by French Catholics. The vote was so close that the academy held a second vote as to whether women in general should ever be admitted. That vote was 90 to 52 against the idea. Indeed, the academy wouldn't admit a woman until 1979.

But that was only the beginning. The worst was still to come.

A short time after the vote, in the spring of 1911, Henri Bourgeois—a newspaper editor as well as the brother-in-law of a woman named Jeanne Langevin, the wife of Marie's lover Paul Langevin—called on Marie with some disturbing news. Jeanne had discovered a trove of intimate correspondence between Marie and Paul Langevin—and she had no qualms about making the letters public.

From most accounts, the revelation was true. Somewhere along the way Marie's close friendship and working relationship with Langevin had blossomed into a full-blown love affair, as their letters attested. In one particularly affectionate note, published by biographer Susan Quinn among others, Marie wrote him that, "It would be so good to gain the freedom to see each other as much as our various occupations permit, to work together, to walk or to travel together, when conditions lend themselves. What couldn't come out of this feeling? I believe that we could derive everything from it: good work in common, a good solid friendship, courage for life, and even beautiful children of love in the most beautiful meaning of the word."

When Langevin in turn poured out his heart about his marital difficulties, Marie shot off a sharp reply displaying an uncharacteristic possessiveness: "But when I know that you are with her, my nights are atrocious, I can't sleep, I manage with great difficulty to sleep two or three hours; I wake up with a sensation of fever and can't work." At least a few circulated letters were even more dramatic. In one, a distraught Marie intimated she might commit suicide if their relationship didn't work out. Irene and Eve "could become orphans between one day and the next if we don't arrive at a stable solution," she wrote. In another, Marie closed by saying, "My Paul, I embrace you with all my tenderness. . . . I will try to return to work even though it is difficult, when the nervous system is so strongly stirred up." When it came to letters written by Paul, fewer exist, although Langevin once wrote that he was drawn to Marie "as to a light . . . and I began to seek from her a little of the tenderness which I missed at home."

It's no surprise that the two were attracted to each other. Langevin had been one of Pierre Curie's star students, and admired the man enormously.

Building on Pierre's early work with crystals, Langevin later would develop an invention that used sonar signals to help Allied military forces detect submerged submarines during World War I. In addition, Marie and Paul had both taught at the Sèvres school for women teachers-in-training. It also hadn't hurt that Langevin, five years her junior, was both handsome and charming.

In scientific circles, it was well known that Paul's marriage to Jeanne Langevin was an unhappy union. Fights between the two were legendary and often involved her abusing him physically. One day he turned up at his lab with bruises; he told his concerned coworkers that his wife, mother-in-law, and sister-in-law had attacked him. On various occasions, Langevin had promised his wife that he would stop seeing Marie. But now Jeanne Langevin possessed the letters proving that their romance was only deepening. And so the warning from Jeanne's brother-in-law was clear: Jeanne was capable of anything, which meant that Marie's life was in danger.

With that, the year turned into one of histrionics, with Jeanne Langevin flying into rages and vowing to rid her family of Marie, whatever it took. One night, when the two women bumped into each other on the street, Jeanne told Marie she'd murder her if she didn't leave France—now. Marie's friend Henriette Perrin later said she'd never forget the image of this illustrious researcher "wandering like a beast being hunted." Yet there were more letters back and forth between the two scientists. By the summer of 1911, Marie and Paul, unable to part ways, were meeting in the Paris apartment Langevin had rented the year before. By this point, he had made a habit of escaping the family home and staying in the apartment for weeks at a time. Eventually he always returned for the sake of the children. The days brought more brawls between Paul and his wife with Jeanne eventually filing charges of abandonment. Soon enough, the French press had gotten hold of intimate letters—or forgeries based on them—and many were published.

Although Marie's daughter Eve, at age six, was too young to understand what was going on, the drama was starting to take its toll on thirteen-year-old Irene, so Marie sent both girls to Poland for the summer. It was their first visit to Marie's homeland. As she had hoped, they loved it.

For all that Marie had going on in her personal and professional life, the writings of Marie and her girls released by Hélène Langevin-Joliot reveal no abdication of her parental duties. No matter how tumultuous the times, Marie always took a few moments to record observations about her daughters' development in her notebook. She recorded when Irene first got her period, stating that "she doesn't lose much blood and scarcely suffers." She recorded personality traits as well. When it came to little Eve, Marie described her as a very sensitive child in tune with the feelings of others, recalling how once when she had "reproached Irene for I don't know what. . . . Eve dissolved into tears." In her letters to her girls, Marie was no gushing mother. But even when the turmoil was at its worst, Marie never lowered the high standards she set for her children. And she always kept a watchful eye. In August 1911, when both girls were in Poland staying with Marie's sister Bronya, Marie wrote to express worry that she hadn't heard from them for a few days. She asked that they write her immediately. In conjunction with her sister, Marie made certain that they enjoyed a lot of time outdoors and also that their intellectual progress was closely monitored. Irene was assigned a half-hour German lesson as well as trigonometry lessons every day of her holiday. Their academic training was evidently rigorous. There's a letter from a defensive Irene even pushing back after being criticized by Marie for her handwriting: "Furthermore, I find that my writing is prettier straight than slanted."

But no matter how busy they were kept, the girls longed for their mother. They often ended their letters to Marie with tender farewells such as: "I kiss you with all my heart on your beautiful tired forehead." Once that summer, when Irene wasn't feeling well, she wrote her mother: "Oh, how I would have liked to have you here while I was sick."

Fortunately, by September, Marie was able to join her daughters in Poland for a happy hiatus—taking the girls on long hiking trips through the mountains—before she had to head off to Brussels and the Solvay Conference in late October.

Almost immediately after receiving that first telegram with news of the Nobel Prize, a second telegram showed up at the conference confirming what Marie had heard earlier. Jeanne Langevin was prepared to release letters to the press proving that her husband was having an affair with Marie. In a politically charged atmosphere that was increasingly intolerant of foreigners—the national elections in 1910 had resulted in a considerable shift to the right in France's Chamber of Deputies—right-wing newspapers stepped up their attacks on Marie even more, tearing apart the legend they had helped create only a few years before. Indeed, many publications took notice of Marie's second Nobel Prize with only a few words on an inside page. As time went on, newspaper after newspaper kept hammering away at the story of Marie's affair but none more so than the ultra-nationalistic *La Libre Presse*, with its chauvinistic catchphrase "France for the French."

Most egregious in the collective mind of the French press was Marie's warning to Langevin in one particularly emotional letter that he not get his wife pregnant during a time of reconciliation. Despite the hardened persona she had always cultivated, Marie showed off her insecurities by telling Langevin that if he resumed sexual relations with his wife—and if she then became pregnant—"it would mean a definite separation between us.... I can risk my life and my position for you, but I could not accept this dishonor." Many in France found this tantamount to treason at a time when the country needed all the offspring it could get to ward off a German threat. As one newspaper worded it, nobody in France should be concerned that Marie might leave the country because of the scandal, but rather everybody should be worried for the "French mother, who ... wants only to keep her children.... It is with this mother, not with the foreign woman, that the public sympathizes.... All French mothers are on the side of the victim and against her persecutors."

A humiliated Marie left the Solvay Conference early and returned to France, where by now the public's animosity toward her was palpable. What should have been a glorious moment—the winning of a second Nobel Prize—had come at one of the worst periods of her life. For the hardworking scientist, the love affair had stymied any celebration and had led, by November 1911, to a quick but ignominious toppling from grace for the world's most famous woman.

Most hurtful was her arrival back at her house near Paris, where she came face-to-face with an angry mob hurling stones at her windows amid shouts of "Go home to Poland," which spoke to the intensity of the public's malice toward her. Marie had no choice but to sweep up her two horrified daughters and seek refuge at the home of her good friends Marguerite and Emile Borel. Earlier, the young Irene, who so idolized her mother, was at school when a friend pointed to a newspaper headline about the Langevin affair. The stunned girl skimmed the story and reportedly burst into tears.

As Eve later wrote in her biography of her mother, people began referring to Marie as a Russian, a German, a Pole, a Jew, or some combination of all four. Mostly, though, she was simply called that "foreign woman" who had come to Paris like a usurper to conquer a high position improperly. Ironically, in earlier years, it was the same scurrilous right-wing tabloid press that had done Marie an inadvertent favor by promoting and thus elevating the public's view of the Nobel Prize, which previously had been scarcely noticed in the field of science. But now, despite their failure to verify the innuendos, hungry reporters seemed determined to topple the icon they had helped build up.

Rather than duck the issue, a protective Paul Langevin challenged Gustave Tery, the editor of a particularly vitriolic newspaper, to a public duel with pistols. Wearing bowler hats and dark suits, the two men faced each other some twenty-five paces apart. Although dueling was still popular, the practices were—by 1911—more of a ritual than anything else, with participants rarely injured. And in this case, too, no one fired a weapon and no one was hurt. After having called Langevin a boor and a coward hiding behind a woman's skirt, Tery ultimately admitted he couldn't raise his weapon against a man who was undoubtedly one of France's greatest scientists. The *Los Angeles Times* thought this event highly amusing, headlining its report: "Pistol Duel Pantomine: Principals Let Seconds Do the Shooting and No Blood Is Let."

Throughout the whole ordeal, Albert Einstein remained supportive of both Langevin and Marie, writing Marie of his admiration for her spirit and energy. (He also admitted to a friend that he didn't believe Marie "attractive enough to become dangerous for anyone.") Pierre's brother Jacques also remained extremely sympathetic and protective. "Doesn't one have the right to sue newspapers for damages?" he asked her. Marie apparently offered no evidence to counter the accusations, although she did at one point issue a statement to the press calling intrusions into her private life "abominable" and vowing to fight for damages. Without providing proof, Paul later charged that the letters had been altered and that certain parts had been omitted. But, for whatever reason, no action against the press from Marie came to pass.

At the same time, Marie also received letters of support from fans in America including the dancer Loie Fuller, who had become acquainted with both Marie and Pierre years before when she contacted them about incorporating luminous radium into her costumes and stage sets. After hearing about the scandal, Fuller wrote Marie: "I love you. I take your two hands in mine and I love you. Pay no attention to the lies, c'est la vie."

But while many friends and colleagues stuck with Curie, some of the most important people in her life did not. One of them was Paul Appell—who had been an extremely close friend and one of her first professors at the Sorbonne. Appell, as dean of the faculty at the Sorbonne, went so far as to start organizing a group of university professors to collectively demand that Marie leave France. When Appell found out that his own daughter, Marguerite Borel, had taken Marie and her daughters in to live with her for a while, he immediately summoned her to his home. Borel found him seething. "Why mix in this affair which doesn't concern you?" he demanded, chucking a shoe against the wall. He told her that the very next day he planned to see Marie so that he could insist, in person, that she move out of the country. In fact, he had already arranged a chair for her in Poland. "Her situation is impossible in Paris. . . . I can't hold back the sea which is drowning her," he said, pushing for a move that Marie never would have wanted.

By her own account, written many years later, Marguerite Borel vowed never to speak to her father again if he yielded to this "idiotic nationalistic movement." Under this threat, Appell caved in and agreed to put off his decision. Borel noted that none of this would have happened if Marie were a man.

By the end of 1911, Eve wrote that her mother had sunk into a debilitating depression. Already she had lost a sister, a mother, a husband, and a father. Three years before Pierre's death, in August 1903, she had also suffered a miscarriage. And now she was about to lose her reputation. There came a point when Marie no longer felt strong enough even to make the journey from work to the house she loved with the large garden in Sceaux, even though it was only about three miles south of Paris. She decided to sell the beloved retreat—located near where Pierre and his mother were buried. It was a place she found hard to part with. But for the sake of her health she bought a flat at 36, quai de Béthune on the fashionable Île Saint-Louis, overlooking the Seine and within easy walking distance of the Sorbonne, a quiet place where she and her daughters intended to live from January 1912.

If things in France weren't good, in Stockholm—where she was due to receive her second Nobel Prize in December 1911—they weren't much better. After news of the love affair broke, a member of the Swedish Academy of Sciences tried to dissuade Marie from traveling to Stockholm to receive her prize in person so that the Swedish king wouldn't have to shake hands with an adulteress. In a letter dated December 1, he cited her published love letters and "the ridiculous duel of M. Langevin" and pointed out in a stinging rebuke that, "If the Academy had believed the letters . . . might be authentic it would not, in all probability, have given you the Prize."

With this, Marie's fighting spirit returned. In an angry letter, she snapped back: "In fact the Prize has been awarded for discovery of Radium and Polonium. I believe that there is no connection between my scientific work and the facts of private life. . . . I cannot accept the idea in principle that the appreciation of the value of scientific work should be influenced by libel and slander concerning private life."

Upon hearing of the exchange, Langevin couldn't stay quiet either. He wrote directly to Svante Arrhenius, one of the men who had nominated Marie: "One cannot judge . . . the correspondence which is reproduced in a distorted fashion—by alterations and omissions . . . if one does not know the condition in which I lived for thirteen years, nor from what kind of people these attacks came."

And so a determined Marie endured a forty-eight-hour train journey to attend the Nobel ceremony scheduled December 10, accompanied by her fourteen-year-old daughter Irene and her sister Bronya. As planned,

King Gustaf V bestowed the prize on her. Although a German newspaper distributed in Sweden had published a report on "The Letters of Marie Curie," she enjoyed a dignified reception from the sophisticated Swedes, especially from the women. In Marie's acceptance speech she acknowledged other scientists who had contributed to the field of radioactivity, but soundly reasserted her claim to be the first to discover its properties. Taking full credit for her accomplishments, she reminded the committee that "isolating radium as a pure salt was undertaken by me alone."

Later, at a formal banquet with King Gustaf, she described radioactivity as "an infant that I saw being born, which I have contributed to raising with all my strength. The child has grown. It has become beautiful."

In her autobiography, Marie wrote nothing of the Langevin scandal, but recorded that the Nobel Prize was a very exceptional honor recognizing her achievements.

The celebration in Stockholm gave way to the ugly reality of Paris. Upon her return, another vicious article deplored the invasion of foreigners at the Sorbonne, charging that the female students were there only to land husbands. In her characteristic way, a strong-willed Marie immediately went back to lecturing at the Sorbonne and working in her small Paris laboratory. But the calm of her routine was about to be shattered again. Less than two weeks after the ceremony, on December 19, 1911, she was buckled over by acute stomach pain and rushed to the hospital on a stretcher with what was thought to be a kidney infection. She recovered from the initial attack, but some old lesions pressing on her kidney—possibly the result of exposure to radiation—required an operation. Marie asked that the surgery not take place before March so that she could attend a meeting of physicists at the end of February.

Although her surgery went better than anyone could have hoped, Marie's health was poor throughout 1912. She grew abnormally thin and often was so weak and feverish that she struggled even to get out of bed. Some thought she suffered from tuberculosis. Others blamed cancer. Secrecy about her illness backfired and sparked a flurry of malicious rumors. The hospital's manager and its Mother Superior warned her of one published article that claimed she was in the hospital about to give birth to Langevin's child. Normally, patient medical records were inviolable, but the pair was so incensed by this libelous charge that they offered to open up her records to the public. Marie agreed. Newspapers revealed her doctor's diagnosis—kidney infection—which mostly stopped the spread of rumors. But what was not mentioned was that she had experienced a nervous breakdown that was the by-product of the deepest, darkest depression of her life, more stubborn than episodes in previous years. Eve later wrote that her mother had grown suicidal. She refused to eat, and her weight dropped from 123 pounds to 103.

After becoming ill, Marie placed her daughters under the care of a Polish governess in early 1912 and barely saw them during the following year and a half, except when they visited her briefly at a house in Brunoy, about thirteen miles from Paris, which she had rented under her sister's name, Bronya Dłuski. A bucolic village, Brunoy was the perfect spot for someone seeking to recover from a traumatic episode. Only a very few close friends and colleagues were told where she was staying. Letters between Marie and her daughters during those first few months show a parent distracted by her own health concerns—but not inattentive. In one, Marie asked Irene if she'd become a bit careless in her cycling excursions. "I'm glad you are starting to become independent, but I do not want us to pay too much for that evolution," she wrote. In another, Marie commented on Irene's grammar, urging her to be more diligent when it came to punctuation. Irene's letters usually included some kind of difficult math equation, which she would have solved perfectly. Marie also wrote often of her concerns over Eve's health after the girl suffered from a variety of mysterious maladies in 1912. In May, she wrote that Eve had a "slight fever climbing to 100 . . . in the evening without other symptoms." By the end of 1912, one of the family's closest friends, the scientist André Debierne, told Marie that her absence was taking a serious toll on Eve's health. Marie managed to pull herself together enough to spend Christmas in Lausanne, Switzerland, with her daughters while Eve recovered.

Marie's seclusion and time away meant she wasn't around to fully appreciate how greatly Pierre's death was continuing to impact Irene. For years, Irene had balked whenever Marie had tried to leave her—even if only for an hour or so. But Marie appeared to have been unmoved. Marie wrote at one stage that Irene "doesn't speak of her father. . . . She no longer seems to be thinking about it, but asked for the picture of her father that we had taken from the window of her bedroom." Yet Irene wrote that she often felt sad and alone. And in a letter from Irene to Marie in August 1909, when Irene was only eleven, she made a point of asking her mother how the plants on her father's grave were faring and which ones were flowering. In a charming note, these were just two of ten very specific questions posed by Irene to her mother. "I ask you ten questions," Irene wrote while on vacation at the seaside. "Respond to all of them when you write to me."

Making Irene's sorrow even worse were the lasting effects of the death of her grandfather, Dr. Eugene Curie, in February 1910. He had been Irene's best friend in the world and both girls' caretaker while their mother spent long hours at work. From her writings, Eve was more in touch with her sister's emotions than was Marie: "I was very young but Irene was [much older]. My grandfather raised her from the time she was a baby and they had a close bond, a great bond. Irene was desolate when first father died and then her beloved grandfather, who had been everything to her."

Irene worshipped the memory of her father and grandfather so it is no surprise that letters and diaries reveal how hard the Langevin scandal was on her. She hated having to hide the Curie name she had been taught to revere and begged Marie to allow Irene to address her letters to Mme Pierre Curie, as usual. But Marie was determined to escape the press in Brunoy and so insisted that her daughter's almost daily correspondence be addressed to Mme Skłodowska.

Although the love affair between Marie and Paul eventually waned, the two remained lifelong friends and associates. After one more separation, Langevin returned to his wife for good, although later he would take on another mistress. (Years later, Langevin would ask Marie to find a position at her radium institute for his mistress, one of his former students named Elaine Montel, and Marie obliged.) The good news for Marie was that a writ of separation between the couple made no mention of her, meaning her name would never be dragged through the sordidness of a divorce court proceeding. In her own comprehensive biography of her mother, Eve never specifically mentioned the Langevin affair, but alluded to this dark period when she wrote about a "perfidious campaign" that led Marie to the brink of suicide and madness. Although Langevin never publicly admitted to having an affair with Marie, his son André later wrote in a biography of his father: "Isn't it natural enough that a few years after Pierre Curie's death this friendship enhanced by mutual admiration should have gradually grown into a passion and resulted in a love affair?"

This storm that erupted in 1911 seemed as if it would never end. Without their precious Mé, as they called their mother, Irene and Eve were despondent. Marie enjoyed a strong will but failing strength. It wasn't clear how, or when, her work and reputation would ever be vindicated, or what would become of the two girls, one of whom was turning out to be exactly like her mother and father in her passion for science. Certainly Marie had veered off the track she'd been taking all her life, since her childhood days in an oppressed Poland.

CHAPTER 2

Moving On

Playtime in the Skłodowski household in Warsaw was exactly what one might expect from a family of educators. When anyone spotted a sunset, Marie's father, Professor Vladislav Skłodowski, would take a few minutes to explain the earth's rotation. When reading a bedtime story in one of the five languages he knew fluently, he would simultaneously translate the work into Polish. Lighthearted banter over the dinner table invariably turned into dense debates over the malignance of inferior educational opportunities for women in Europe. A caring man who taught physics, science, and mathematics, there's no doubt that the Polish patriot valued a love of learning more than anything else in the world. As such, no matter what else was going on, he never missed an opportunity to impart some of his own knowledge about scientific matters to his children.

Skłodowski became the primary caregiver to ten-year-old Marie and her three older siblings after his wife, also an educator as well as a gifted piano player, died of tuberculosis in 1878 at the age of forty-two. (Another child, a girl, had died from typhus at the age of twelve a few years before her mother's death.) Marie later called the loss of her mother the first great sorrow of her life, one that threw her into a profound depression. She wrote that her mother's influence on her was extraordinary "for in me the natural love of the little girl for her mother was united with a passionate admiration."

Skłodowski, too, was heartbroken over the loss of his beautiful wife and daughter. He vowed to devote himself to his work and to the education of his remaining children. As long as he lived, he believed his offspring capable of great things—especially when it came to his little "Manya." Once, when she was only eleven, the headmistress of her school told Skłodowski that, although Marie was at the top of her class, she was much more sensitive than her peers. Perhaps, she suggested, he should consider holding the girl back a year. Skłodowski did exactly the opposite. He immediately pulled Marie from the school's nurturing environment and enrolled her in a Russian-run school that catered to high achievers. A Skłodowski was always supposed to go forward, not back. Even later, when as a teenager Marie was taking a break from school and was employed as a governess in the countryside, Skłodowski never failed to include increasingly challenging math problems for her to solve along with every letter he sent.

Skłodowski had every expectation that the brightest of his brood would go on to be somebody special—and Marie didn't disappoint. As the star pupil, she was awarded a gold medal at her high school graduation in 1883. She was always at the top of her class. Since women weren't welcome at the University of Warsaw, Skłodowski encouraged Marie and her sister Bronya to join a brave circle of bright friends in attending the so-called Floating University, an illegal night school that constantly changed locations in order to evade the watchful eyes of the Russian authorities.

During Marie's youth, her father wove the ideals that would shape her future: a love of country, a love of learning, and a love of humanity. With war clouds on the horizon—and many heartaches ahead—her childhood of hard work, academic rigor, and high expectations would prove to be a perfect training ground for what was to come.

In her writings, Marie often spoke of the love and respect she felt for her father. She said that, even as an adult, she couldn't go back to Poland without meeting people who held "tender memories" of her father. In particular, she never forgot how her father had been deprived of a laboratory by his Russian taskmasters. Often he had referred to the overseeing of a major institute of research as the dream of a lifetime. That dream was shared by Marie's husband, Pierre—and then by Marie herself. No matter what life threw at her, the drive to reach this goal would never go away.

But in the aftermath of the Langevin scandal, and with war rumors starting to dominate people's lives, Marie and her daughters would have much to get through before coming close to achieving this hope, this vision, this fantasy.

First and foremost was the matter of Marie's health. After winning her second Nobel Prize in 1911, Marie was often away, scouring the countryside for one health cure or another—always far from the crowded confines of Paris. Paying the price were Irene and Eve, shuffled endlessly among governesses and various family members, forever homesick for their mother's embrace.

By the summer of 1912, Marie's health had deteriorated to the point that she was taken to a sanitarium in the Savoie mountains, part of the French Alps. At the time, all sorts of people visited sanitariums, believing cold clear mountain air to be the best remedy for almost anything that ailed them. Strangely, perhaps, Irene's letters to her mother indicate that the girls celebrated Bastille Day on July 14, 1912, with Paul Langevin. Apparently Irene and Eve maintained a warm relationship with the scientist—just as their mother did—despite all that had transpired. In later years, neither Irene nor Eve ever wrote anything specific about the Langevin scandal, so it isn't clear what they thought about their mother's alleged actions, although they always remained close to Langevin.

What is clear is that Irene hated the idea of hiding her family name more and more, especially as her mother continued traveling incognito from one health center to another in order to stave off publicity. Feeling nostalgic for her father, a defiant Irene asked if she could spend time with Pierre's brother. Marie agreed, and Irene and Eve went to stay with Jacques Curie and his extended family in Montpellier. Marie wrote that Irene "adores everything relating to the Curie family: the name that she bears and that she would not like to change—that of the family of her father."

A constant flow of letters between Marie and her girls during this period reveals Irene growing in strength and maturity, with none of the broodiness typically displayed by teenage girls. Meanwhile, a much more emotional Eve seems to have been deteriorating in health and spirit without a parent around to nurture her.

A keenly intelligent Irene wrote daily to her mother, often to ask how to solve a certain kind of math problem. Indeed, math was brought up in almost every letter exchanged between the two just as it had been in letters between Marie and her father, with long complicated equations often included as part of the correspondence. In one letter, Irene, obviously advanced for a fourteen-year-old girl, remarked that "the derivatives are coming along all right, the inverse functions are adorable. On the other hand, I can feel my hair stand on end when I think of the theorem of Rolle, and Thomas's formula."

But a thoughtful Irene also asked often about politics, saying that there were many issues she didn't understand, such as the causes and consequences of the Turkish government. Overall, her letters display a maturity beyond her years and an overwhelming desire to please her mother. In one, Irene makes a point of emphasizing how hard she was working not to confuse her languages while simultaneously reading works by Charles Dickens, Gotthold Ephraim Lessing, and Shakespeare. "I have a dictionary on my table at all times," Irene wrote her mother.

But when Irene was with Pierre's brother, she also expressed a youthful exuberance: "At Montpellier, I met uncle. In the morning we walked in the town. In the afternoon we were in Palavas on the edge of the sea. We were together, uncle and I, all day long. Uncle is so happy to have me here."

It was during her weeks in Montpellier, in 1912, that Irene first began referring to her mother as "Ma chérie" or "My dearest," a salutation indicating Irene and Marie to be on more equal footing than the previous greeting of "Ma douce Mé" or "My sweet Ma." But Irene would always look up to her adored mother and, in almost every letter, she expressed concern over her mother's health.

On the other hand, Eve—although only seven years old—already showed that she cared very little for academic pursuits. Her letters were often more carefree, no more than one or two sentences in length, and always ended with an expression of how much she missed her mother.

Differences between the two sisters were readily apparent even in the most trivial of situations. Often, Marie sent her girls small boxes of candy. For her part, Eve gobbled up every piece by the very next day. Irene, on the other hand, savored her bits one by one, hiding them away in a drawer. Eve would come across her sister's stash, hardly touched, some time later. Much to Eve's chagrin, a disciplined Irene had no problem rationing her supply to a piece a day. When it came to money, too, Eve immediately spent whatever she could get her hands on while Irene hung on to every last franc simply because she could think of nothing she wanted to buy.

In the summer of 1912, Marie was well enough to leave the sanitarium. But she wasn't yet ready to return home. She accepted an invitation from one of the few good female friends she would have in her life—the well-known British physicist Hertha Ayrton—to visit her in Hampshire on the edge of the New Forest in England. This time—finally—the change of scenery worked to get the ailing Marie moving again.

Perhaps the visit reaped rewards simply because Marie liked Ayrton so very much. They had first met in 1903 when Marie took a trip to London with Pierre and had remained in touch ever since. Like Marie, Ayrton was not only the widow of a distinguished scientist but was also an important practicing researcher herself. As soon as Ayrton had read about the Langevin affair, she immediately wrote to Marie, offering her rented home at the seashore as a sanctuary. In addition to being a brilliant physicist, Ayrton was also a crusader for women's rights, and she knew how to deal with both the press and the public. She promised Marie complete privacy—and Marie was not disappointed.

The two women soon were joined by Marie's daughters and their Polish governess after Irene celebrated her fifteenth birthday in Brittany—without her mother—on September 12.

Ayrton would tutor Irene on the side, treating her as a serious student and discussing math with her on an adult level. Irene was delighted. She won over Eve, too, by accompanying the budding eight-year-old musician on the piano, even when she played French songs. More than anything, though, she piqued the older girl's interest in women's rights. Earlier that year, an astute Irene had written her mother, "I've noticed that every day or almost every day an English minister just misses being killed by the English suffragettes, but it seems to me that that isn't a very brilliant way for the suffragettes to prove that they are capable of voting." But Irene grew more understanding after Ayrton shared with her the British way of dealing with women who fought for their rights: they threw them in prison. Up to that point, Marie had always been reticent to lend her name to causes, believing scientists should remain objective in all political matters. But Marie thought so highly of Ayrton that she agreed to allow her to add her name to a petition protesting the imprisonment of leaders of the suffrage movement.

Following her return to Paris in late 1912, Marie finally began feeling more like her old self, both mentally and physically, and so, as 1913 got underway, she began writing more about her work and less about her health. After an absence of more than a year, she resumed making notes in her lab book again in December 1912. She also went back to teaching classes at the Sorbonne. Maybe it was just the passage of time, but it seems that she began rediscovering her inherent drive and common sense. Any lasting longings for Langevin seem to have been brushed aside as there exists no further evidence from this time onward of correspondence between Marie and Paul of an intimate, personal nature.

Back with her girls, Marie behaved very much like her father when it came to the day-to-day nitty-gritty of running a household. She never permitted loud voices in her home, whether angry or happy ones, just as they had never been allowed in the Skłodowski home of Marie's youth. Like her parents and siblings, Marie and her own daughters handled their average lives with composure even in the most trying of circumstances. No matter what, one's studies were of utmost importance but, at the same time, physical activity was never discounted. Hélène Langevin-Joliot said that, "Marie never liked to hear children being loud and there was never any screaming that occurred." She also noted that there was never any question in either household over whether a woman was equal with a man. Both could and should enjoy the same aspirations.

To Marie's great satisfaction, Irene at least, so much like Pierre, was turning out to be a brilliant student with a penchant for science. She wanted to make certain she received every advantage possible.

Therefore, as early as 1910, despite the eruptions in her personal life, Marie had joined forces with her lover, as well as a number of other eminent French scholars, and organized a private educational cooperative to take over the teaching of their children. It was an early—and unusual—flirtation with homeschooling in a country that has always prided itself on the quality of its highly centralized school system. Despite Marie's proclivity for hard work, she despised the French educational system for bogging children down too long in the classroom. Marie's wish was for her children to have more physical exercise in the fresh air as well as practical experience so that they might grow into more well-rounded adults.

Through the cooperative, each member contributed to educating everyone else's children—including Irene—in their respective homes. The curriculum was varied and included not only scientific topics but also such diverse subjects as Chinese and sculpture. In the mornings, the cooperative's mobile class of colleagues' children filed into a laboratory at the Sorbonne to hear the great Jean Perrin explain chemical processes. Afterwards, Paul Langevin would go over mathematical problems or Marie would recite the principles of physics. All the while, the children had plenty of opportunity to enjoy physical exercise. In their free time, "we did gymnastics, swimming, bicycling, horseback riding . . . we rowed, we skated," Irene later wrote. During holidays, the girls even learned to ski before it became a trendy, mainstream sport. The cooperative provided Irene with a remarkable education filled with instruction from some of the most brilliant minds in France. It continued until Irene turned fourteen. After that, Marie felt her daughter needed a more traditional education and she enrolled her in a private school, the Collège Sévigné, where she would soon earn her degree.

Marie's mindfulness of her children's education is one reason she refused when a delegation of Polish professors headed by writer Henryk Sienkiewicz, the most celebrated man in Poland, had urged her to leave France and return to Poland in 1912. Following the Langevin scandal, many colleagues in France would have encouraged such a move. But Marie knew her children's futures would be more promising if they remained in Paris. Not that she didn't listen intently to what members of the delegation had to say. As a student she had always imagined that one day she would return to Poland. And no doubt she was intrigued by the delegation's ambitious plans for her to oversee a radium institute that the Scientific Society of Warsaw hoped to build for her. The idea of making a contribution to the place of her birth appealed to her sense of duty. But she just couldn't take her girls away from Paris and the Sorbonne. Not wanting to let her countrymen down, though, Marie offered instead to send two of her brightest Polish assistants to run the institute in Warsaw under her direction from Paris. The group agreed to her offer as did the two young Polish assistants.

Although Marie was often away, there were a few memorable holidays worth noting during this period including at least one with Albert Einstein and his family. In July 1913, the Curies and the girls' governess met up with the Einsteins and their eldest son, ten-year-old Hans Albert, for two weeks of hiking in the Alps. The Curie girls got along famously with Hans and they became lifelong friends. Eve later said that she "was amused at the way Einstein circulated absentmindedly among the boulders, so deep in conversation that he walked alongside deep crevasses and toiled up the steep rocks without noticing them." Less than a year and a half after meeting Marie for the first time at the Solvay Conference, Einstein and his wife had been invited to stay with the Curies in March 1912 while Einstein was addressing the French Society of Physicists at Langevin's invitation.

About two months after the trip with the Einsteins, on September 12, 1913, Marie missed Irene's sixteenth birthday because she was attending a conference in England. She wrote her daughter that she would have loved to be with her. "I hope you get this so you know I don't forget you. . . . I cover you with kisses," Marie wrote. By this time, missing her daughter's birthdays had become something of a habit for Marie. Such familial milestones seemed to exist only in the margins of her consciousness.

Being mostly in Paris, Marie and her girls wouldn't correspond regularly again until the summer of 1914, a time when, as Irene noted, the international situation had become extremely "disquieting" and the girls were separated from their mother. By then, Irene had passed both parts of the baccalaureate exam and was planning to enter the Sorbonne in October. But first there was their annual vacation in L'Arcouest in

Brittany, where Irene had gone with Eve, a cook, and the governess to lay the groundwork for her mother's arrival. According to letters collected by Langevin-Joliot, Irene was by this time dispensing advice in the voice of a responsible caregiver. "If you feel that you need to take care of yourself," she wrote her mother, "don't hesitate. I beg you to delay your arrival to effect your cure." She also kept her mother abreast of her sister's educational development: "Eve works a lot. She doesn't want to do arithmetic, but one shouldn't bother her about that because she puts really a lot of good will into doing other things, even German." Increasingly, Marie and her daughter were becoming more like partners. "I feel already," Marie wrote Irene, "how much you have become a companion and friend to me."

Irene would have had to grow up fast anyway. A major world event was on the way that summer, one that would cement Irene's position as a dedicated scientist in her own right, a woman with her own unique set of exceptional talents to offer.

The summer of 1914 was as eventful as they come. As the major powers of Europe were plotting a series of diplomatic maneuverings designed to weaken the Kingdom of Serbia—which would ultimately lead to war—Marie was happily putting the finishing touches on her dream of a lifetime: the construction of her highly anticipated Radium Institute in Paris, one that would be dedicated to Pierre's memory. The institute was to be concentrated entirely on radioactivity and would be built with the help of the Pasteur Institute and the Sorbonne.

Only three years earlier, fellow scientists had openly questioned whether Marie should even be allowed to remain in the country. But now, feeling better than ever and with her reputation back intact, Marie was tending to the final details of what she hoped would be a place of research envied around the world. Marie was intimately involved in every decision related to the institute. To ensure that workers would gaze out on lovely gardens, Marie spent hours putting a variety of flowers in the ground herself. From the start, Marie agreed to divide the institute into two sections. One section, devoted to research in physics and chemistry, would be headed by Marie while the other, devoted to research in medicine and biology, would be headed by Claude Regaud, one of France's leading scientists. At long last, the building was about to open on Rue Pierre Curie. Marie was thrilled.

But the preamble to the European war, starting with the assassination of Archduke Franz Ferdinand of Austria at Sarajevo on June 28, 1914, got in the way of Marie's best-laid plans.

For some time, those in France had been forewarned that a war might be coming, but many doubted that the rumors would come to fruition.

But, on August 3, just two days after declaring war on Russia, Germany declared war on France. The occupation of Paris—once unthinkable—suddenly seemed highly plausible. From L'Arcouest, Irene wrote that she refused to cry or panic "because that will just panic the others." In Paris, Marie had hoped to wind up her work so that she could join her daughters at the seaside. But then she had a change of heart as conditions worsened. "Things seem to be going badly . . . we are awaiting the mobilization momentarily," she wrote Irene and Eve. "Be calm and courageous. If my departure becomes impossible, I will stay here and you will return as soon as possible."

In the end Marie believed it wasn't wise to abandon her nearly operational Radium Institute, even if all of her able-bodied male workers had been called to serve the war effort. Most especially, Marie wasn't about to leave behind her radium supply.

With the German army advancing toward Paris, the French president moved the government's headquarters to Bordeaux. He then gave Marie the key task of transporting the nation's supply of radium there as well, fearful it might fall into German hands. She packed up a suitcase filled with tubes of radium bromide encased in lead—a suitcase so heavy people still aren't sure how she carried it—and boarded a train to Bordeaux. The trip took ten hours because the train stopped so often. Not wanting to be recognized or hassled, she waited until everyone left the station's platform before getting off the train. She was met by a government representative who took her to a meager private home because all the hotels were filled with fleeing Parisians. The suitcase never left the foot of her bed. The next day, the representative took her and her suitcase to the University of Bordeaux, where the radium was stored in a vault. She returned immediately to a nearly deserted Paris on board a train packed with soldiers. One of them asked: "Are you not Madame Curie?" Wanting to be left alone, she replied, as she often did, "No, you are mistaken."

Of course, Marie also fretted over the fate of her family after Poland was invaded by Austria on August 12. She worried over the fate of her friends after Germany took control of the French city of Amiens on August 30. And most especially, she was also genuinely concerned about her daughters. Indeed, this is one of the few times on record when Marie, forty-six at this time, poured out motherly affection for her children without restraint.

"I am dying to come and hug you. I don't have the time. . . . There are moments when I don't know what to do. I want to hold you close so badly," Marie wrote in one letter. Wanting to ease her mother's worries, Irene wrote to Marie every single day, even when she had almost nothing to say. Without fail, she always pleaded to return home. "We must be together in this time of trial," Irene wrote her mother. It was a difficult separation especially after some locals accused Irene and Eve of being spies when they were caught speaking Polish to the help. With the Germans gaining ground, Paul Langevin enlisted as an army sergeant while other scientists were called to the front. Marie's close friend Marguerite Borel took over the running of a hospital. Her friend Hertha Ayrton in England invented a kind of fan designed to blow gases from the trenches. When the government held a nationwide drive for gold and silver, Marie offered her Nobel medals. But the Banque de France declined to accept them. Marie was desperate to find a way to offer her services to her adopted country. And, soon enough, she found it.

After seeing so many soldiers arriving back in Paris, their bodies destroyed by unnecessary probing, she knew there was a real need for X-ray equipment and technicians out in the field. With the help of X-rays, doctors could determine the exact location of bullets, shrapnel, and bone fractures without having to perform exploratory surgery.

At the start of the war, the Military Board of Health had no authoritative body of radiology. Radiologic installations existed in only a small number of major hospitals—and certainly not out on the battlefields. Determined to fill this void, Marie immediately gathered together all the apparatus she could find. She set out in August and September 1914 to create several stations of radiology located as close to the soldiers as possible; the operation of the stations was ensured by volunteers Marie had trained herself. With the help of the Red Cross, she also fitted up a radiologic car, or a touring motor car, that quickly transported radiologic equipment to doctors in the battlefields. Soon this one car turned into a fleet of cars despite the challenges thrown at the effort by government officials. For example, four months after receiving permission for a second car, an ornery official told Marie she most likely wouldn't be granted a permit to travel with a radiology car due to stricter regulations slapped on women. In the end, though, the military was unable to stop Marie and before long she established about two hundred permanent radiology posts and eighteen radiology cars to service them.

The Battle of the Marne—an Allied victory on September 12, 1914, that effectively ended the month-long German offensive that had reached the outskirts of Paris—convinced Marie that it was okay to send for her daughters. They were ecstatic at the idea of returning home. Eve went back to school, and Irene began studying for her baccalaureate in mathematics and physics at the Sorbonne, as well as taking a nursing class.

As months dragged on, Marie found herself in dire need of more technicians in the field. And so, in 1916, Irene's education was interrupted so that the teenage student could act as her mother's assistant. Initially, she accompanied her mother to the battlefields but within a few months Irene was left alone to look after a field radiological facility in Belgium. Working within earshot of gunfire, Irene exuded a calm inherited from her parents. She thrived in the wartime environment and on her new responsibilities. Irene, by herself, X-rayed the wounded and told the surgeons exactly where to probe. And she taught herself how to repair equipment. Irene didn't even mind spending her eighteenth birthday in 1915 training nurses to take her place so that she could move on to another battlefield position. Rather, she relished the chance. She wrote proudly to her mother that on that birthday she had located four large shell fragments that were successfully removed from a soldier's hand. She ended, "I spent my birthday admirably . . . except that you weren't there."

Somehow, the two kept up their good humor during this time. Irene wrote that she had to stifle a laugh as she once watched her mother profusely thank a government inspector for his kindness in authorizing the installation of new radiologic operations. She was well aware that this same man and others had made Marie's work extremely difficult, brandishing regulations as they took their time in approving her requests for visas and requisitions.

For four years, Marie crisscrossed the country, taking her radiologic mission to battlefield after battlefield, often with Irene at her side. Thanks to Marie's fleet of what would eventually be twenty ambulances, called "Les Petites Curies," more than a million wounded men were provided with X-ray services, undoubtedly saving countless lives. She was incredibly proud of Irene who had shown staggering courage in difficult circumstances. These years marked some of the most memorable times for mother and daughter. During the war, Irene had worked not only at the front between Furnes and Ypres, but also at Amiens, two places bombed by aircraft.

This was obviously a tense time for Eve, who was mostly away from her mother and sister. In July 1918, Eve was staying with friends at L'Arcouest while Marie was in Paris. Eve wrote that she was very frightened to hear about a German air raid alert the week before. "In any case, it reassures me a little to know that you went down into the cellar, at least during the worst part, and I hope that you will do the same again," she wrote. Eve went on to explain that she'd received an account of the alert from a friend who also told her that Notre Dame might be in flames. "This is by way of saying that all of this does not reassure me one bit," wrote Eve, who added that she had little to do but exercise and pursue her musical interests.

The next month, in August 1918, Irene wrote Marie of her delight that France was finally on the offensive and was "full of hope." She said that "the Germans are at last surprised on their side. I think, my dear, that we are finally arriving at the moment we have been waiting for, the moment where—after having gone through the worst—we are coming out of it." She noted that the lowest point for her had come during the German offensive against the English. Irene must have by this time joined her sister because she next wrote on August 25 about going canoeing and swimming with friends at L'Arcouest, saying that the tide had turned with reason finally triumphing over ignorance.

Indeed, only three months later, on November 11, 1918, Marie and Irene were working side by side when they heard sounds that were a far cry from those of bombings: they were the festive sounds of cheering, cannon fire, and church bells. After four years, the war was over. But 10 million were dead.

At the end of the war, Irene received a military medal for her efforts. Perhaps mindful of how Pierre had turned down its offers of accolades when he was alive, the French government presented nothing to Marie. Marie later wrote of these times: "Of the hospital life of those years, we keep many a remembrance, my daughter and I. Traveling conditions were extraordinarily difficult; we were often not sure of being able to press forward, to say nothing of the uncertainty of finding lodgings and food."

Marie also wrote that nothing was so moving as to be with the wounded and to take care of them, but it was an experience that made her and Irene forever hate war.

With the war over, Marie could finally turn her attention to her new laboratory. She wanted desperately to get back to creating the best possible research institute not only to honor the memory of both her father and Pierre, but also so that her skillful Irene would have a place to one day perform world-class work.

But, following years of war, there was one major stumbling block: a lack of funding. By 1919, she had next to no equipment except for two measuring instruments she was able to buy thanks to a small grant from the National Research Fund. France, unlike other countries, simply had never funneled adequate money into science—and certainly wasn't able to do so following a devastating war. And, at fifty-two, Marie didn't have a huge income; she was supporting herself and her daughters on her salary as a university professor, a salary depressed by rampant inflation.

Marie paid little attention to money, another characteristic she had inherited from her father. When Pierre was alive, both he and Marie had categorically refused to patent their discoveries, even though funds from patents would have helped finance further research and discovery. Although they did obtain some early financial support, Pierre in particular had considered himself an independent outsider, one who perhaps let his pride prevent him from doing the kind of networking that might have led to more funding opportunities. Both Marie and Pierre were dedicated to a higher cause and were part of a community of disinterested researchers. The question often came up of whether or not Marie and Pierre should apply for a patent for the process to produce radium. They were always adamantly against the idea. One of their mutual overriding beliefs was that pure research should be carried out for its own sake and must not become tangled up with industry's profit motives. Researchers, they felt, should selflessly make their findings available to anyone and everyone. As a result, Marie and Pierre were exceedingly generous in supplying their fellow researchers, competitors included, with the preparations they had so laboriously discovered and developed. They furnished industry, too, with everything they knew.

And that was just one reason it was so difficult for Marie to live out the dream of building a first-class research institute from scratch after the war.

She had received a substantial grant many years earlier, in the winter of 1907, from a great admirer, the American philanthropist Andrew Carnegie. This had enabled her to at least begin to assemble a staff of promising scientists who could devote themselves full-time to research. Carnegie, who oversaw the expansion of the American steel industry in the late 1800s, was a big admirer of Curie's work, inviting her to visit him in Pittsburgh in November 1906. (She didn't go.) As the years went on, she always kept an eye out for Polish and women applicants with potential when handpicking the so-called Curie scholars funded by Carnegie. But the only gram of radium Marie had at this point—barely a thimbleful—was in constant demand from doctors and medical researchers. Still stinging from the way she had been treated by the press during the Langevin affair, Marie refused to publicize her need for funding. It seemed that she was stymied—but only temporarily.

Fortunately, about this time, an American came to the rescue—a petite spitfire named Missy Meloney.

CHAPTER 3

Meeting Missy

It was a bright morning in May 1920 and Mrs. William Brown Meloney, nicknamed Missy, could not have been more nervous as she waited to be called into the interview. Usually she had no fear of speaking with any man or woman, no matter how famous they were. In the past, the accomplished journalist had won an audience with the likes of Benito Mussolini—on four separate occasions—as well as with Adolf Hitler and others known around the world. But an exclusive with the great Madame Curie was the interview she most wanted.

This moment had been years in the making. Earlier, Meloney had approached a top French newspaper editor for advice on how to reach Marie. Not surprisingly, he told her that the reclusive researcher would see no reporter as she absolutely abhorred publicity, preferring instead to stay in the background. He was right. Stung by the hyperbolic judgments of the press during the Langevin scandal, Marie avoided the media at all costs. Yet, time and again, an undeterred Meloney would request a meeting, only to receive nothing back—not even a written acknowledgment of the solicitation. But then she caught a break. A French novelist and mutual acquaintance named Henri-Pierre Roché—impressed by Meloney's tenaciousness and even more by her charm—stepped in and personally delivered a final handwritten letter of appeal: "My father, who was a medical man, used to say that it was impossible to exaggerate the unimportance of people. But you have been important to me for twenty years, and I want to see you [for] a few minutes." Something about that simple letter touched Marie, and the busy physicist finally agreed to see Meloney—for a few minutes.

Neither woman could have imagined where those few minutes would take them.

Eleven years younger than Marie, who was fifty-two at the time, Meloney was born in Kentucky in 1878. She came from good stock, raised by a woman of remarkable determination. Not only was Meloney's mother a college graduate, but she also established a school for freed black slaves in 1876. Later, as a widow, she was named president of the Washington College for Girls and also wrote for a women's magazine. As a child, Meloney had planned to be a concert pianist, a dream that was shattered by a horseback riding accident at the age of fifteen that caused her to suffer health problems and forever walk with a limp. Meloney once declared: "I have been lame since fifteen, and had a bad lung since seventeen and have done the work of three men ever since." At seventeen, she followed in her mother's footsteps and became a full-time reporter for the *Washington Post*. Later she married a newspaper editor, William Brown Meloney, and they had one son.

Meloney took ten years off to raise her son, but then jumped back into the journalistic arena with a job as bureau chief for the *Denver Post*, becoming the first woman to get a seat in the US Senate press gallery. That led to her meteoric rise to the helm of *The Delineator*, one of six major women's magazines that published fiction, articles on fashion, and tips on home decorating. But it also published major pieces on the world's leading movers and shakers. When she met up with Marie, Meloney was in Europe as editor of *The Delineator* on a visiting tour that included meetings with H. G. Wells, J. M. Barrie, and Bertrand Russell. But it was the interview with Marie—her longtime idol—that mattered more than the rest of them put together.

Meloney met Marie at her Radium Institute, inside a badly decorated office that, for some reason, Meloney later wrote "might have been furnished by Grand Rapids, Michigan." Before Meloney could spout out a single question, Marie shocked her interviewer with the news that

America possessed about fifty grams of radium while she—who had worked for years to discover the element—possessed no more than one gram in the safe of her poorly equipped facility. This set Meloney on a course that would alter the fate of physics and rally American women toward a common goal.

Often described as looking like regular table salt, radium is a weighty metal with some real staying power: one gram will decay to half a gram in 1,580 years. But it is prohibitively time consuming and costly to separate from other substances. In the late 1800s, Marie and her husband Pierre, were forced to sift through tons of an ore called pitchblende in order to prepare just one gram of radium, a hard, tedious, and dangerous process that stretched across four years, from 1894 to 1898. By 1920, the element had become something of a commodity, albeit an expensive one, transferred from sites of industrial production to those of widespread consumption among the masses. In particular, the couple's pioneering work on how the radiation from radium could damage the tissue of living organisms—a discovery that was soon tested and put to use against cancer and other illnesses—was vitally important for the medical community.

The Curies intentionally decided not to patent the process to extract and purify radium, but it wasn't surprising that others soon began seeking to cash in on the element's perceived benefits. For example, after a radium-bearing ore called carnotite was discovered in Colorado, extraction plants sprang up in or near Denver, Pittsburgh, and New York City with the help of the US Bureau of Mines and several corporations. Before long, the wealthy United States became the biggest radium-producing country on the planet, churning out 80 percent of the world's production. Between 1912 and 1922, the United States produced more than 170 grams with the help of some of the finest physicists and chemists in the country.

Marie's difficulties arose from the fact that she simply didn't have the backing or the finances or the manpower to access the radium that was necessary to perform any new experiments. Her own country, suffering from serious coal and food shortages after the war, was in no position to channel a large amount of money into scientific research. A lack of office staff meant Marie was forced to reply to a nonstop flow of mail herself. There was no way she would ever be able to achieve her dream of transforming her Radium Institute into a world-class facility. Even the one gram Marie possessed had been earmarked not for scientific research but for the medical community seeking to treat cancer.

When Marie told Meloney that the going rate for a gram of radium was between $100,000 and $120,000, a price way out of the scientist's reach, the journalist was aghast. She asked Marie why she had never patented the rights to radium production, which would have given her more than enough money to buy the radium herself. But Marie insisted again, as she always had, that radium was an element belonging to all people and that it was not meant to enrich any one person.

Meloney was astonished. She knew full well the impressiveness of both the laboratories in America and the factories where radium ore was treated in mass from having visited them personally. To her, it was mindboggling that the two-time Nobel laureate was conducting some of the world's most important research in a deficient lab and coming home each evening to a simple apartment that had the requisite furnishings but nothing even remotely hinting at the importance of its occupant. The government had provided Marie with a small pension following the war, but it was mostly eaten up by widespread inflation. Indeed, it was the war's destruction of prime agricultural land that led to an increased need for imports, meaning prices for flour and other staples had risen dramatically. On a positive note, since 1919, the US Army had paid Marie $75 a week to teach military men waiting to return to America how to use X-ray equipment.

When Meloney visited Marie at her apartment, she later wrote, she was stunned that Marie greeted her at the front door herself because she couldn't afford a maid. Thomas Edison, after all, had a lab packed with every manner of contemporary equipment. So did Alexander Graham Bell, whose family enjoyed a grand estate with many fine horses. Meloney was touched by Marie's modesty and by the way she had shunned material goods. As Hélène Langevin-Joliot put it, Marie "was convinced science was good for all of humanity and not just for herself or for her own personal gain and it was as simple as that . . . and this was something that moved [Meloney] greatly."

Marie's challenges, Meloney thought to herself, were more than just a hook for her article. They were her call to arms, so to speak, one that she was more than willing—on the spot—to take up and get behind. Here, before her, was her hero, a woman of profound abilities who never sought any financial gain for herself. Making herself vulnerable, Marie had laid bare her problems for this stranger to see. It was more than Meloney could stand. She must, and she would, make sure Marie obtained her radium and whatever else she desired. Those who knew Meloney wouldn't have been surprised. Her keenness always trumped everyone else's and she was, if nothing else, a woman who could make things happen. After all, her network of contacts reached all the way to the White House, where even Vice President Calvin Coolidge was a friend.

But whether Marie knew it at the time, bringing the scientist to America always would be, for Meloney, an essential part of their future collaboration. Indeed, their connection would push the envelope for both women. Their exchange proved a pivotal moment, filling Meloney with a new sense of purpose and Marie with a new source of funds and good will. Most importantly, a beautiful, lifelong friendship was born.

Before Marie came to know Meloney, Marie's sister Bronya had been her main confidante. She traded many loving letters with other family members as well as with her childhood friend Kazia, but through her adult years Marie had lacked the time and even the desire to form a lot of female friendships. She wrote that when she was collaborating with Pierre "it can be easily understood that there was no place in our life for worldly relations. We saw but a few friends, scientific workers, like ourselves . . . we also maintained affectionate relations with my husband's brother and his family." She called this mode of "quiet living" necessary in order to achieve "the great work of our lives, work begun about the end of 1897 and lasting for many years."

As Langevin-Joliot said, "Among women scientists, she had a few good friends but not many. . . . Mrs. Meloney was one of the very few friends she ever had." Perhaps because of the humiliations she had endured during the Langevin scandal, Marie was particularly susceptible, at this stage in her life, to Meloney's flattery and idolization.

But almost immediately it became apparent the two fast friends wouldn't always be on the same page. Unlike Marie, Meloney was a social conservative who believed women should not work outside the home when their children were young. She told her readers that it had pained Marie enormously to spend so much time away from her daughters when, in fact, Marie had made her work a priority and didn't seem to feel conflicted by her choice.

Although Langevin-Joliot said Marie's feelings for her daughters were "enhanced by her husband's death," others might have argued that Marie's lengthy separations from Eve, in particular, came close to neglect when her girls were quite young. Marie's research always took precedence. In a letter written in June 1920 by fifteen-year-old Eve to Marie—when Eve was away at the beach and her mother was working in Paris—Eve complains of being poorly prepared for her baccalaureate exam after having received encouragement only from a family friend. "I'm just going to try my luck but I have very little hope of being accepted," she said. In a sweet notation, an obviously eager-to-please Eve tells Marie that "for Friday" she planned to learn: "optical, electricity, magnetism, induction, descriptive, algebra, Julius Cesar, Dickens, all the literature." She ends the list by saying: "And I know nothing. . . ." As always, she tells her mother how much she misses her.

But perhaps Meloney chose not to fully recognize all these separations. Once, the journalist sent Marie a novel called *This Freedom* about children whose lives are ruined by a mother working outside the home. Marie wrote Meloney that she disagreed with the book, saying that "I agree, of course, that it is not easy for a woman to bring up children and to work out of the home. But I don't believe either that the author is right in his conclusion. . . . I don't think that he has considered the rich women who leave the children to a governess and give most of their time to social visits and fashion."

Meloney glossed over any obvious differences of opinion in the April 1921 issue of *The Delineator*, which was devoted almost entirely to Marie. On the cover was the drawing of a gorgeous young lady with the tagline: "A Beautiful American Working Girl." Underneath was written "The Greatest Woman in the World: Madame Curie." The lead editorial,

"That Thousands Shall Not Die," was splashed above the table of contents. Articles in the issue ranged from one titled "Do You Fear Cancer?" to another titled "How Do You Wash Dishes?" In many of her articles on Marie, Meloney emphasized the scientist's "sacrifice" at having had to be separated so often from her daughters, especially during times of crisis. Meloney wrote: "She felt she had to apologize for being absent from her children, even in the war." She also described Marie as a "woman of rare beauty" with a face bearing "the mother look."

Meloney was savvy enough to know that her campaign to help Marie would only work if she was able to perpetuate the image of a poor woman who had forfeited even time with her own children in order to benefit the greater good. And Meloney seems to have hammered away at this theme despite objections from Marie. In a letter to Meloney in November 1920, Marie bluntly tells her that several lines in one particular article written about her were "not entirely justified." Although she wasn't rich, Marie told her friend, neither were other French scientists. "I don't complain . . . or feel unhappy about it," she said. Yet Marie was a somewhat complicit partner as she always expressed gratitude to Meloney for any and all efforts that would lead to her receiving some radium.

In the months after that first meeting between Meloney and Marie, Meloney used her bully pulpit to galvanize a nation of women into action—women who were feeling especially powerful after having just won the right to vote in 1920. Seemingly overnight, the Marie Curie Radium Campaign was created to begin gathering donations for Marie. Meloney was no stranger to hard work, and asking people for money had become almost second nature for her. Already, from her position as editor, she had led a successful relief campaign for postwar Europe. In addition, under Meloney's stewardship, *The Delineator* had initiated the Better Homes Movement, a nationwide campaign of home ownership, modernization, and beautification undertaken after World War I in response to a critical shortage of homes. Local campaigns were initiated to encourage people to own, build, remodel, and improve their homes, with advice on home furnishings and decorations distributed.

Meloney put that savvy organizational skill to good use with her new pet project. She organized an advisory committee of scientists, including the president of the American Medical Association. As the appeal was to be especially one from women to women, she also formed another all-female advisory committee, a group that included Mrs. John D. Rockefeller, Mrs. Calvin Coolidge, and Mrs. Robert Mead, founder of the American Society for the Control of Cancer. Soon there were two committees in the mix, one male and one female, generally referred to as the Marie Curie Radium Committees. At first, Meloney's goal was to get ten ultra-rich women to donate $10,000 apiece but, much to her chagrin, not one of those contacted was able to contribute such a large amount of money. Meloney decided to change tactics. Instead of asking for sizeable sums, she went to her readers seeking $1 to $5 donations.

It worked. Her message resonated with an admiring public and spread to all sectors of society. Immediately, housewives went door to door, pleading Marie's case. Inspired college women launched collections on their campuses. Feminists who felt Marie was the perfect representative for their modern movement ran with the cause. Even children, who heard of the campaign from their mothers, saved their nickels and dimes. As Eve later wrote of the effort, "nothing is impossible in America."

And this was all before Marie ever set one foot on American soil.

As her crusade made gains, Meloney kept perpetuating the image of a scientist so unselfish she now had no choice but to turn to a sympathetic American public to go on with her work. And her work, Meloney insisted, was nothing less than finding a cure for cancer. She wrote to her readers: "And life is passing and the great Curie getting older, and the world losing, God alone knows, what great secret. And millions are dying of cancer every year!"

Marie had made clear to Meloney that her contribution to the cancer effort was an indirect one—but it seems Meloney was oblivious. She also had told Meloney she'd be using any radium received from America for scientific purposes—not medical ones. Indeed, the gift would replace the radium she had isolated years before, which was now being utilized for cancer therapy. In the nineteenth century cancer was often treated by radical surgery. But after Marie's discovery of radium, doctors began placing radiation in small containers as close as possible to the tumor site. Before the war, Marie's involvement with radium therapy was primarily

through her lab's supply of the international standard against which measurements were applied. But during the war she helped prepare radon
vials for the treatment of wounded soldiers in military hospitals. Then,
after the war, a permanent department for radium therapy was set up at
Marie's struggling Radium Institute. But Marie's top priority always had
been, and continued to be, pure scientific research.

Whatever liberties Meloney took, they succeeded. Less than a year
after her first meeting with Madame Curie, the Marie Curie Radium
Campaign had raised $156,413 for the purchase of a gram of radium
as well as equipment for Marie's lab. An exuberant Meloney wrote to
Marie, "The money has been found; the radium is yours." The campaign
awarded a contract to the Standard Chemical Company of Pittsburgh,
Pennsylvania, for the gram of radium. At the same time, Meloney also
arranged for an American publisher, the Macmillan Company, to publish Marie's biography of her husband, giving Marie a source of royalty
income over the next several years. Macmillan published *Pierre Curie*,
which also included a brief autobiography by Marie, in November 1923.
As Langevin-Joliot said, "I wouldn't be surprised if at least a bit of the
autobiography was written by Mrs. Meloney. . . . I certainly think some
of the words and passages were thought of by Mrs. Meloney." Indeed,
the first eighteen pages of *Pierre Curie* are a tribute to Marie written
by Meloney in which she says that "When most of us shall have been
forgotten, when ever the Great World War shall have dwindled to a few
pages in the history books, when Governments shall have fallen and risen
and fallen again, the work of Marie Curie will be remembered."

But when it came to the money raised in America, there was a long
string attached. In exchange for the outpouring of help, American
women—led by Meloney—wanted Marie to pay a visit to America.
This placed Marie in a quandary. She had always run away from the
limelight. The public attention of a trip to America absolutely mortified her. Surely Marie felt tempted to go to the radium-rich America
where, no doubt, she could personally thank her benefactors. And, after all, America had taken on new significance in the eyes of many
Europeans after the war; it was a land that seemed more youthful and
more energetic compared with a region still reeling from industrial

destruction and massive civilian casualties. But Marie also felt that the idea of weeks on the road and meeting so many new people was too overwhelming to even contemplate.

In January 1921, Marie wrote Meloney to say that if she were to come to America, she would stay no longer than fifteen days.

But Meloney had other ideas.

Again, Meloney whipped into action. She promised to limit Marie's public appearances. She promised to go to Europe and make the trip across the Atlantic with her. Most significantly, she promised to muzzle the American press so that no reporter would make mention of the Langevin scandal. From a purely selfish point of view, Meloney also realized that resurrecting the love affair could backfire and set back the fundraising effort. America in the 1920s was more puritanical than France had ever been, so she had no problem promising that this delicate subject would remain under wraps. Eventually, Meloney cajoled pledges to suppress the Langevin story out of every editor she approached, more than once managing to also charm a contribution to the Marie Curie Radium Campaign from them during the same conversation.

But even as time went on, there was a huge disconnect between how Marie envisioned a tour of the United States and how Meloney saw it. If she agreed to the journey at all, Marie didn't want to come until October 1921. Meloney wanted, instead, May and June. Surely, Meloney told her friend, Marie didn't want to miss out on receiving all those honorary doctorates she was likely to get at the long list of commencement exercises she was lining up. Marie wanted to stay no more than two weeks and insisted it would be "difficult to accept giving lectures, except for those which are absolutely necessary." Meloney imagined, and had promised, a much longer itinerary packed with innumerable appearances, including the ultimate spectacle that was to be the presentation of the gram of radium at the White House.

The spitfire Meloney—who had crafted a well-oiled public relations machine surrounding the scientist and a possible US tour—specified exactly how things were to be and just assumed Marie would eventually go along with her plans. No matter what the objection, Meloney dismissed it. Meloney wrote Marie: "You say you don't want to leave your daughters? We invite your daughters too. Ceremonies tire you? We shall draw up the most reasonable and limited program of receptions. Come! We shall make it a fine journey for you, and the gram of radium will be solemnly presented to you at the White House by the President of the United States in person."

Marie was touched by her friend's can-do spirit. She finally relented, agreeing to a full seven-week stint beginning in May 1921—but not without some real apprehensions. To collect her gram of radium and to thank her lovely new American friends for it she would have to conquer her fears and accept for the first time in her life, at the age of fifty-three, the obligations of a great official journey. She'd also have to give up time at work. "Marie was so happy in her lab . . . she did not want to lose time," Langevin-Joliot said. Two months before the trip, in March 1921, Marie confided her misgivings to *Le Matin*'s editor Stéphane Lauzanne, one of those who had encouraged her to go. He later wrote about how "this great woman—the greatest woman in France—was speaking haltingly, tremblingly, almost like a little girl. She, who handles daily a particle of radium more dangerous than lightning, was afraid when confronted by the necessity of appearing before the public."

Marie's trepidation was no match for the enthusiasm of her daughters, who were delighted with the prospect of a new adventure in an exciting America alongside their mother. The fashion-conscious Eve even convinced the unstylish Marie to buy a new dress. In the last ten years, Marie had so often been away from home that seven weeks of uninterrupted time together helped strengthen the family bonds. Eve would later describe her childhood as an unhappy one. She had a melancholy sense of the years slipping away without giving her the chance to really get to know her mother. Her letters indicate that Eve was resentful of the time her sister shared with Marie, especially when the two traveled the country during the war. When Eve was sixteen, Marie wrote Irene after a summer holiday in Brittany, "I hope . . . our Evette will love us more in Paris than she did in L'Arcouest." Later she also wrote Irene that, "We will have to reconcile the scientific work which we two represent with the musical art represented by Evette, which is much easier to do in good weather than in rain."

There's no doubt Marie had grown closer over the years to Irene, who was so much like Pierre in temperament and intelligence. At one point, just before the American trip, Marie was so worried about Irene's passing a French exam required in order to teach in the state-controlled educational system that she wrote the vice rector of the Sorbonne, as well as the minister of education, asking that special consideration be given to Irene's contributions during the war. A curious exchange of letters reveals a very protective mother taking great pains to point out that the only reason Irene might be behind in her studies was that she had devoted two years to national defense. Like Marie, Irene was growing passionate about scientific research, and nothing was more important than her education.

As the day of departure grew closer, countless letters began pouring in from America, asking Marie to lecture, endorse this product or that, make a public appearance, or simply to hand out advice. Marie's sense of dread grew. She culled through the letters with the help of Dr. Robert Abbe of New York City, a plastic surgery pioneer who had introduced radiation therapy to the United States. Abbe, who helped Meloney in her fund-raising campaign, had visited the Curies as early as 1903 after corresponding with them about their research on radiation. Abbe transported some samples back to the United States and began experimenting on cancer patients. A huge fan of Marie's in particular, in later years he created a garden at his summer home in Bar Harbor, Maine, in which two swans—named Pierre and Marie—floated in the pool. During her visit to America, Marie was to attend a gala hosted by Abbe at the College of Physicians of Philadelphia. Grateful for his support, Marie would bring Abbe a souvenir: a piezoelectric apparatus used by Pierre. As a young man, Pierre—along with his brother—had discovered an instrument called the piezoelectric quartz electrometer, which was able to measure faint electric currents. Abbe later wrote in a book titled *Mme Curie* that it was the privilege of American women to lay the tribute of radium at her feet—a tribute with which Marie "can and will give back to them a thousand fold more in value."

In the weeks preceding the trip, many newspapers covered the nonstop string of tributes and ceremonies being planned for Marie in seemingly every corner of the United States. Everyone was trying to figure out what new distinctions they could bestow upon the scientist so that she might leave the country with all the official titles her stellar reputation deserved. It was hardly comprehensible to the American press and public that, in France, Marie was not a part of the prestigious French Academy of Sciences, which had denied her a seat in the election a decade earlier. It was also surprising that she did not possess the Cross of the Legion of Honor.

An editorial in the French periodical *Je Sais Tout* also declared it disgraceful that other nations should honor Marie while her adopted country did so little. Faced with so much criticism, and seeking to make up for lost time, a few French officials suddenly took great pains to claim Marie as its special native daughter, especially after finding out that the American president himself planned to make the presentation of the gift of radium. The government hastily offered her the Cross of the Legion of Honor. But—as was standard form—she refused it, as had Pierre in 1903. (Pierre had thanked the government, but argued that he'd much rather be awarded an adequate laboratory than any sort of decoration.) Marie was years later to ask that the rank of Chevalier of the French Legion of Honor be presented to Meloney.

Eventually, at the urging of *Je Sais Tout,* the government planned a farewell celebration in Marie's honor at the Paris Grand Opera on April 27, 1921, just a week before she was to set sail, that was characterized as a benefit for the Radium Institute. As a result, Marie agreed to take part, attending along with her daughters and Meloney, who by this point had made her way over to Paris. The gala was glamorous, marked by a reading from the great actress Sarah Bernhardt titled "An Ode to Madame Curie." In attendance were some of France's most prominent personalities, including President Aristide Briand, who went on to win the Nobel Peace Prize in 1926. Marie's close friend Jean Perrin described Marie's contributions and the promises her discoveries had rendered possible in the years ahead. Unfortunately, though, Marie had a hard time following what was going on, despite wearing new eyeglasses with stronger lenses. She hadn't wanted to tell anyone about the constant buzzing in her ears or about her failing eyesight, which she privately feared might be related to her exposure to radium. She had, however, confided in Meloney, who had scheduled a visit with a leading eye specialist in Manhattan.

Even so, it was quite a sendoff for the woman who had never been fully appreciated by her own country. Everything was in order. And now, finally, the Curie women—with Missy at the helm—were ready to set sail for the United States of America.

CHAPTER 4

Finally, America

On May 11, 1921, the RMS *Olympic* steamed into New York Harbor at the mouth of the Hudson River. Nicknamed "Old Reliable," the 882-foot vessel was the lead ship of the Olympic-class liners built for the White Star Line, which had also included the *Titanic*. Marie and her daughters stood on deck alongside Meloney and watched the emerging shapes and details on land growing steadily more distinct. With nearly three times as many people as Paris, New York City in the 1920s was emerging as the financial and cultural capital of the world, with an unrivalled skyline. Each week, vast numbers of immigrants entered into that same harbor, one of the busiest in the world, feeding the thriving economy. And the city's women were becoming more powerful as well. Wearing lighter and shorter clothes, they drank and danced and, as of the year before, they also enjoyed the right to vote.

Marie wasn't doing any of these things back in France, where women wouldn't win the right to vote until 1944. And she had never been politically active even though Irene, with leftist tendencies, would later become involved in social movements.

In exchange for funding, though, Marie would have to allow herself to be held up by those in America struggling for women's liberation as living proof that women were every bit as smart and capable as men. She also would have to allow herself to be ogled at every turn. When Albert

Einstein, the other global scientific celebrity, arrived in New York for his first visit to America the month before, he was greeted by the same sort of hysteria that would greet the Beatles four decades later. Marie's visit was a similar sort of extravaganza, unique in the history of science, that would inspire a new wave of fans—but this time most of them were women.

On the deck, dressed simply in her conservative black dress, Marie faced her first American-style press conference. Earlier, Meloney had warned Marie that although the usual customs formalities would be waived, easing their departure off the ship, the pier would be packed. And it was. The lively crowd of fans that had been waiting more than five hours included hundreds of women from the Polish American Society carrying roses. A large contingent of Girl Scouts, as well as doctors and nurses, waved welcome banners. Esteemed members of various scientific committees joined a handful of New York Camp Fire Girls, who brought Marie a handsome red leather bag beaded with a fleur-de-lis design. About a dozen others waited patiently to present Marie with larger bouquets of flowers, along with expressions of admiration. Mingling among them all were countless fans brandishing American, French, and Polish flags—all hoping to catch a glimpse of this epitome of power for modern women—as brass bands played three national anthems at the same time.

But it was the pack of photographers and reporters, all jockeying for position, that terrified her most. Although the word "paparazzi" wouldn't make its way into the American lexicon until it was used in Federico Fellini's film *La Dolce Vita* in 1960, no one at the time knew better than the American press that customers would pay a premium to read about celebrities. And, in America at least, Marie was a celebrity—a big one. In France, Marie enjoyed a placid and modest life. Although she almost always refused interviews, if she ever did agree to speak with a reporter, it was almost always during a brief one-on-one exchange over a cup of tea. As Eve later wrote, her mother in France "had succeeded in convincing her compatriots, and even those who came nearest her, that a great scientist was not an important personage."

This moment in America would be different. After taking away her black taffeta hat and handbag, Meloney settled a befuddled Marie into a comfortable armchair on the boat deck where she watched as Meloney corralled more than forty journalists into a sloppy but still menacing semicircle around her. In rapid succession, they shouted out one question after another—asking her what she planned to use the radium for—while commanding her to look this way or that. Marie answered as best she could, mostly in the monosyllabic manner of a reserved scientist, while quietly offering her thanks to the American people. She said she'd always wanted to bring her daughters to "this great country." Before she knew it, all sorts of strangers were shaking her hand so hard that her arm started to ache. Irene and Eve, twenty-three and sixteen at the time, looked on from the sidelines, bemused at the sight of their mother as the object of such veneration.

Whether Marie liked it or not, the frenzy of that hour-long circus was just a brief taste of the seven-week endurance test ahead. How would she get through it? The reluctant hero had no idea. Yet this new land of promise she'd heard so much about was the source of everything she needed right now. After so much time apart from the girls throughout their lives, she took heart in the company of her daughters, who were about to have the time of their lives.

On that first day in America, even after having survived the realities of a week-long, numbingly cold 3,000-mile crossing of the Atlantic, the fifty-three-year-old was judged by a *New York Times* reporter to be "energy personified . . . walking with a quick step. Several of those at the pier spoke of the clearness of her eyes, and the lively interest she took in everything she saw." Marie had spent most of her trip from France holed up in her cabin, a bridal suite arranged for her by Meloney. No doubt Marie's quick gait had more to do with wanting to flee from the mob as fast as possible than with any sort of laudable energy level.

From the ocean liner, the Curie women were whisked away after the press conference in a limousine sent by Andrew Carnegie's widow to Meloney's Greenwich Village home at 31 West Twelfth Street. Marie thought it charming. Built originally in two parts in 1895 and 1900 by the department store owner George A. Hearn—Macy's chief rival—the Beaux-Arts building boasted a rock-face brownstone base with elegant iron balconies extending across the front. To get inside, the women had to negotiate an obstacle course of roses laid out near the entrance by a horticulturist whose cancer had been cured by radium. In previous days, all sorts of floral bouquets had been delivered to the house, most of which Meloney sent on to a children's hospital. Certainly Marie wasn't used to receiving gifts and being the object of so much adoration. "Even before she got to America the people there were captivated by the idea of [Madame Curie] and all that she stood for," said Hélène Langevin-Joliot.

Following a restful night's sleep, Marie awoke to a disturbing headline in the *New York Times*. Above the front-page article describing Marie's arrival the day before were the bold words "Mme Curie Plans to End All Cancers." Marie fumed at this misrepresentation of her intentions, one that was printed again in the body of the story. "Radium is a positive cure for cancer," the *New York Times* reporter quoted Marie as saying. "It has already cured all kinds of cancers, even deep-rooted cases. . . . Those [doctors] who have failed do not understand the methods." Exactly what Marie had said has never been made clear, but she had never before made the claim that she could cure all cancers. If anything, the scientist with no formal medical training had always tried to emphasize that radium might not be an absolute remedy for every type of cancer. But Meloney herself had perpetuated the myth that Marie's work was the answer to prayers with a recent editorial in *The Delineator* titled "That Thousands Shall Not Die," which played up the idea that Marie's work was treating cancer.

No doubt the so-called magical properties of radium had excited many a physician by the early 1920s. In discovering radium, both Marie and Pierre also discovered that its radiations produced powerful physiological effects. (Pierre, for example, found out the hard way that radiation seriously burned tissue by trying it on his own skin.) But although Marie had worked with doctors toward the goal of using radiation to reduce or even completely decimate tumors, she knew that radiation alone wasn't always sufficient in the permanent eradication of a malignancy. Sometimes, it had to be combined with surgery or other treatments. And she also knew that radiation could kill healthy cells. Even so, this therapy known as Curietherapy in France, and radiation therapy elsewhere, is still used in many instances to treat cancer today.

When it came to the *New York Times* article, Meloney assured her troubled guest that the mistake would be corrected and immediately jumped on the phone to make sure it was. The paper that had run the article on the front page ran a correction the very next day—on page sixteen.

That problem settled, another challenge quickly cropped up. When discussing the various university diplomas that would be bestowed upon Marie during the trip, Meloney offhandedly asked her, "Naturally you've brought your cap and gown? They are indispensable for these kinds of ceremonies." Of course Marie hadn't brought a university gown; she didn't even own one. As the only female professor, she never had a need for such a garment at the Sorbonne. The enterprising Meloney again sprang into action, hastily recruiting a tailor to craft something appropriate out of black silk and velvet. Again, though, a modest Marie thought the whole exercise silly and unnecessary. And she refused throughout the entire seven weeks to wear the cap, arguing that it looked ridiculous and wouldn't stay affixed to the top of her head anyway.

Following a welcome lunch at the home of Mrs. Andrew Carnegie at Ninety-first Street and Fifth Avenue, one that was attended by Mrs. Cornelius Vanderbilt and socialite members of the fundraising committee set up by Meloney, the Curie women enjoyed a brief tour of New York City before setting off in a coach for their lecture circuit to women's colleges across the Northeast. First up was a visit on May 13 to Smith College in the lively town of Northampton, Massachusetts, where Marie was made an honorary doctor of science. There, Marie was greeted by hundreds of exuberant young women, strikingly presented in white dresses tied up in colorful sashes. The first woman scientist to win worldwide acclaim was walking on their campus. The students could hardly believe it. Almost any student at any women's college at the time would have felt a kinship with such an accomplished woman.

Smith president William Neilson acted as Marie's escort as they viewed the gymnasium, a classroom and, most importantly, the chemistry lab, where Marie inspected the students' work. As they walked around, Meloney trailed behind, carrying Marie's flowers for her. Afterwards, in John M. Greene Hall, Marie sat perched high on a platform looking uncomfortable as she faced some two thousand people and listened quietly as one professor after another stepped up to verbalize the value of her research. In a low voice, Marie briefly expressed her appreciation to the college, in English, and thanked Neilson for his kindness. She also took a moment to respond to a story one of the professors had just told to illustrate Marie's habit of becoming completely lost in her work. He had described an incident when Marie was engrossed in her lab research and a maid had rushed in crying out that she had swallowed a pin. Marie replied, "Never mind, here's another." Before the amused crowd, a deadpan Marie agreed that it was a good story, but that "unfortunately, though, it never happened."

During an informal reception in the library, a dozen or so reporters tried to get close to Marie but were blocked by Meloney, Marie's daughters, and others. Later, Marie would write how exhilarating it was to interact with so many fresh-faced female students. And yet reporters kept at bay while covering the tour consistently described the guest of honor as looking ill, feeble, and exhausted.

At one point during the lecture tour, a reporter noted that Marie seemed dazed. According to his description: "Her arms hung lifelessly, her features were ashen gray, contrasting strongly with the simple black tailored suit she was wearing. She smiled from time to time wanly and patiently, and there was an occasional gleam in her blue-gray eyes. The deep lines on her face, however, showed how seriously the unaccustomed strain of her whirlwind visit to America has affected her." Another reporter also expressed concern over Marie's stamina, writing that: "Doubt has been expressed by several who saw Mme Curie yesterday of her ability to complete this itinerary without a nervous breakdown." He went on: "A quarter of a century of painstaking research work in the field of chemistry has left the world's foremost woman scientist in complete possession of all her keen faculties. The first two hectic days of her two weeks' tour [it would really be seven] of this country, following immediately after a seasick voyage [she was never seasick] from France, have transformed her into a listless being, tired nigh to death, and obviously ill."

But yet the tour forged ahead. From Northampton, Marie and her entourage went by automobile through the Berkshire Hills, an excursion that took them past a string of charming New England villages. In one, South Hadley, Massachusetts, she was greeted by crowds of adoring spectators lining the roads, mostly students at Mount Holyoke who knew from all the publicity that she'd be traveling their way after her visit to Smith. Finally, they pulled into Poughkeepsie, New York, where Marie was scheduled to spend two full days at Vassar College. Marie was immediately struck by the beauty of the college's 1,000-acre campus on the banks of the Hudson River. And she was impressed that, unlike some other women's colleges, Vassar offered young women a full range of courses, from art history to zoology, all taught by leading scholars. The next day, on May 14, Marie delivered one of her most substantial speeches as she discussed the unexpected consequences of pure research. "When radium was discovered, no one knew that it would prove useful in hospitals. The work was one of pure science. And this is a proof that scientific work must not be considered from the point of view of the direct usefulness of it," she told the crowd. "It must be done for itself, for the beauty of science, and then there is always the chance that a scientific discovery may become like radium a benefit for humanity."

By this time, doctors such as Marie's American friend Robert Abbe had touted success after using sealed sources of radiation to reduce the size of tumors whether in the breast or elsewhere in the body.

If anything, the speech underscored Marie's overwhelming commitment to science for science's sake—and not for the express benefit of mankind. It was evidence of how the reality of Marie clashed—again—with Meloney's portrayal of her.

Marie ended by encouraging the students to carry on with their scientific work and to keep "for your ambition the determination to make a permanent contribution to science."

Although the *Springfield Republican* newspaper kept predicting that a worn-out Marie would quit the tour early, Eve later wrote that Marie had learned a great deal during this portion of the trip. Obviously she basked in the respect of the students, who'd all given at least a dollar to her fundraising campaign. But she also gleaned a lot about America's university life—so different from what she knew back home. She couldn't get over the greenness of the grounds, the modernity of the facilities, the cleanliness of the dormitories, and especially the inclusion of physical activity into daily schedules. "The students play tennis and baseball; they have gymnasium, canoeing, swimming, and horseback riding," Marie later wrote. Having had to scrimp to go to college, Marie said she was deeply moved by what she perceived to be a very democratic system of education, one in which wealthy students took classes alongside poorer students able to attend college on scholarships.

The personalities of the students, too, made an impression. Marie told the *Herald Tribune* that the young American women were very different from French students. "Over there all are sad and most dressed in black from the war," she said. "I have been strongly impressed by the joy of life animating from these young girls." The spirited coeds she encountered couldn't have been more different from the youthful Marie, who had been a somber student so engrossed in her studies she sometimes forgot to eat.

Overall, these days Marie spent at women's colleges were busy and frenetic, but they were also manageable and irresistibly interesting. To her credit, Meloney always carved out some time for Marie to enjoy a bit of peace and quiet, as when the party stopped on its way from Smith to Vassar beside a stream in the Berkshires for a mellow picnic lunch. On another occasion, Meloney spontaneously arranged one night for Marie to accompany a Vassar physics major named Margaret Hill to the laboratory. Hill was honored to show Marie how she conducted various experiments with the help of a new Curie electroscope, developed by Pierre decades earlier to measure radioactivity. "I was so excited and awed by Madame Curie's coming to Vassar that I could hardly stand it," Margaret Hill Payor wrote seventy years later.

Back in New York, Marie and her girls spent time resting at Meloney's home before making a heavily publicized appearance as the guests of honor at an event on May 17 hosted by the American Chemical Society at the Waldorf-Astoria. As early as 1909, the American Chemical Society had elected to make Marie an honorary member, and she smiled as she entered the dining room, greeted by several minutes of applause. Dr. Francis Wood, the head of the Crocker Cancer Research Laboratory at Columbia University, told the six hundred or so people in the audience—who had paid $5 each to attend—that Marie had done more to help humankind than any other scientist of her time.

Playing up Meloney's recurring theme, he emphasized Marie's human-itarianism more than her scientific acumen. Following several speeches during which she was praised repeatedly, Marie was presented to the au-dience as "queen of the scientists of the world." She received a standing ovation lasting several minutes. But then she promptly sat down without formally addressing the crowd. Quite frankly, her energy was spent.

Wood was one of Marie's biggest fans, and a good friend, and had done much to help Meloney raise money. His emphasis on Marie's hu-manitarianism would not go unnoticed, at least by some women scien-tists. Christine Ladd Franklin, a Columbia University instructor, argued in the *New York Times* that Marie should be seen above all else as an example of what women could accomplish in science. She lamented the fact that, during her American tour, no one ever made mention of the "Curie" as being the unit of radioactivity when, she argued, other distin-guished scientists had been lauded for having their names attached to the designation of specific units such as "watt" and "volt."

By this time unsteady on her feet, Marie barely managed to muddle through brief stops at receptions hosted by the National Academy of Sciences, the American Museum of Natural History, and the New York Mineralogical Club after leaving the Waldorf-Astoria.

But the tributes would only grow more grandiose. The next day, on May 18, Marie stepped on stage at Carnegie Hall to applause from 3,500 members of the International Federation of University Women that carried on for several minutes. Among the invited guests to this ma-jor event—hailed as the largest gathering of university women ever held in America—were the French and Polish ambassadors and Marie's old friend, the former Polish Prime Minister Ignace Paderewski, who looked on proudly as Marie garnered still more titles and medals, including the Ellen Swallow Richards Memorial Prize worth $2,000. Dr. Florence Sabin of Johns Hopkins Medical School saluted Marie for proving that a woman could be accomplished in her career while, at the same time, per-forming the duties of a devoted wife and mother. Others including the highly respected Dr. Alice Hamilton of the Harvard Medical School also sang her praises. How different were these lengthy tributes from what the reclusive Marie was accustomed to in France!

But the most notable speech of the day came from Dr. M. Carey Thomas, president of Bryn Mawr College, herself a pioneer educator and suffragist, who used the occasion of Marie's visit to lay out an ambitious women's political agenda: educational reform, the abolition of prostitution, and equal pay for equal work. Now that they'd won the right to vote, she said, women were seeking "to act as we think best . . . the right to dispose of our own lives and bodies . . . to live worthily and unashamed." Thomas, who had also played an active role in raising money for Marie's radium fund, concluded with a bold plea for women to organize nationally and internationally and, under threat of revolution and overthrow, to force governments to disarm.

In comparison, Marie then stood up timidly and, in a hushed voice, delivered just twenty-seven words that could scarcely be heard except by the dozen or so people sitting closest to her. She said, "I thank you from the bottom of my heart for the welcome you have extended to me, and I shall never forget the warmth of your reception."

Marie is not known to have shown any particular interest in the women's suffrage movement, either in America or in France, and it's not clear what she thought about Thomas's call to arms. When asked about her grandmother's views, Langevin-Joliot said that although Marie wasn't overtly involved, "it happened once that her name was cited in a discussion at the French Parliament. . . . [I]t was stated that she was against allowing women to vote. Marie Curie protested in newspapers that this was not at all her opinion, and that she thought women must be considered as citizens and allowed to vote." But Marie had never called herself a feminist and had always been too busy in the lab to become politically active. Perhaps she was perplexed at being the trigger for Thomas's call to action.

Even so, the gatherings of so many people hammering away at the issue of women's rights didn't go unnoticed. Later, after she had returned to France, Marie told an audience that "the men, in America, approve and encourage the aspirations of women." No doubt her interactions with so many ambitious women would have led her to that conclusion. But, unbeknown to Marie, there were several times during her American tour when men had expressed discomfort with—or downright hostility toward—her and her accolades. In one case, the Yale scientist Bertram Boltwood had written his friend Ernest Rutherford in Europe that, when he learned "the Madame" wished to call on him in New Haven, Connecticut, he immediately reached out to Yale authorities to argue that he "had no desire to have the honor thrust upon me and that I considered that it was the duty of the institution to entertain her." When he learned that Yale, "on the recommendation of a couple of medical men," had voted to give her an honorary degree, he told them that they had been "a little hasty in their action." Marie heard nothing about all this. She visited Boltwood, and he was "quite pleasantly surprised to find that she was quite keen about scientific matters and in an unusually amiable mood. . . . She certainly made a good clean-up over here. . . . But I felt sorry for the poor old girl, she was a distinctly pathetic figure. She was very modest and unassuming, and she seemed frightened by all the fuss the people made over her."

Boltwood, who was belligerent toward Jews as well as women, noted that he was hugely relieved Yale hadn't awarded Einstein a degree when he visited in April 1921. "Thank heaven. . . . We escaped that by a narrow margin," he said. "If he had been over here as a scientist and not as a Zionist it would have been entirely appropriate, but under the circumstances I think it would have been a mistake." (Einstein went on to win the Nobel Prize in Physics that year.) Marie, ignorant of Boltwood's true feelings, later referred to her visit to his laboratory as one of the highlights of the entire trip. Perhaps she'd forgotten that Boltwood was one who opposed allowing her to prepare the international radium standard when both attended the Solvay Conference some ten years previously, in 1911.

Sentiments similar to Boltwood's also may have been behind a private vote by Harvard's Physics Department not to award an honorary degree to Marie. When a dismayed Meloney dogged the retired Harvard president Charles Eliot for the reasons behind the decision, he replied that the physicists believed that the credit for the discovery of radium should not have gone entirely to her and that, what's more, she had done nothing of major importance since her husband's death in 1906. Meloney's reply in defense of her friend was admirable but not enough to change anyone's mind. "[T]he outstanding virtue of these years," she wrote Eliot, "lies in the fact that having discovered radium and come into prominence she turned to her home as a normal mother and gave the intimate, minute attention to her children which motherhood should impose." Again, Meloney emphasized Marie's supposed ability to combine work and family. Decades later, in 1955, Harvard would finally award an honorary degree to a woman—Helen Keller.

In general, at least some scientists felt hostile to Marie's fundraising tour not because she was a woman, but for financial reasons, arguing that money given to Marie's Radium Institute was money their own laboratories weren't going to get. But there were a few who seemed genuinely flummoxed by the idea of a woman ever doing the same kind of work as a man. An editorial published in the *New York Times* after Marie's arrival acknowledged that she was an example of a woman attaining prominence in a scientific domain. Yet "the majority of women are still to develop either the scientific or the mechanical mind," it said.

After Marie offered her quiet thanks at Carnegie Hall, the country's most promising female science students formed a line in front of her. One by one, they each bowed and handed her a flower—first a rose, then a lily, followed by an orchid, and so on. The record-breaking celebration was topped off with the fifty-voice-strong Vassar choir singing "The Star Spangled Banner."

Watching Marie, a reporter judged her to be shy, weary, and somewhat bored. He was undoubtedly right. By the time she arrived at Carnegie Hall, she had grown tired and disinterested, just sitting and listening all the time, and her aching right arm was in a sling. At least one doctor prescribed complete rest. But Marie and Meloney had decided anyway to squeeze in a surprise visit to Hunter College in New York City on May 18, where they toured the laboratory. The occasion was a forty-five-minute speech on the theory of radioactivity delivered by Irene in perfect English before one thousand students in the chapel. Throughout the trip, Marie prevailed upon both of her daughters time and again to represent her at receptions and other gatherings. The two proved a dynamic duo: Eve delighted in the fancy parties and charmed the guests with witty small talk while Irene, though hardly at ease with strangers, still made a competent substitute by speaking authoritatively like the bright young scientist she had become. The press seemed enamored with both girls but for once paid more attention to the friendlier Eve. One reporter dubbed the beautiful teenager—who, unlike her sister, always dressed in the latest fashions—"the girl with the radium eyes." Not one prone to jealousy, Irene ceded the limelight to her younger sister.

But the main event was still to come. No matter how tired Marie felt, she had come so far and there was no way she was going to miss her trip to the White House. Finally, she was going to get what she came for. All these public appearances—and all this learning about how to play the American system with Meloney's help—were finally going to pay off.

CHAPTER 5

The White House

With the passage of days, the physical and emotional toll of the journey—and especially of standing up to the scrutiny of so many strangers—had mounted steadily. Even before coming to America, Marie, at fifty-three, might have been taken for a woman in her late sixties. Now she looked even more doddering. But at long last, this woman was finally going to get what she was paying a high personal price for— her precious radium.

After arriving in Washington on May 19, the night before the main event, Meloney sat down with her antsy friend to explain the logistics of what was about to happen. At a very formal 4 P.M. ceremony, President Harding would lead Marie, Irene, and Eve into the East Room of the White House and, following brief remarks, present Marie with the deed to the radium, inscribed on a scroll tied up in red, white, and blue ribbons. Next, the president would hand her the gold key to a specially constructed lead-lined mahogany casket weighing approximately 130 pounds—displayed on a nearby table but with nothing inside. Indeed, unbeknownst to most of those who would be attending the ceremony, the radium itself was being safeguarded at the nearby US Bureau of Standards, where officials had tested it, measured it, packaged it, and agreed to deliver it safely to Marie's ship just before she set sail for France the next month.

By this time, even Meloney had to glean from Marie's appearance that her growing list of ailments was a lot more serious than anything that could have been easily attributable to age or regular fatigue. Her health complaints included anemia, dizziness, blurred eyesight, kidney infection, sudden drops in blood pressure, and a continuous ringing in her ears. Although not a doctor, Marie apparently knew what was happening to her, whether she admitted it to anyone or not. Marie later wrote that "my work with radium . . . especially during the war, had so damaged my health as to make it impossible for me to see many of the laboratories and colleges [in America] in which I have a genuine interest." But American doctors, perhaps swept up in Meloney's enthusiasm over the scientist's visit, initially failed even after thoroughly examining her to acknowledge any connection between radium and Marie's debilitating condition. "There is nothing the matter with Mme Curie at all," insisted Dr. Edward Rogers, "except that she has been trying to do too much." He conceded that Marie was somewhat anemic—but not more so than any other person working intensely in a constrictive environment with little free time—and maintained that there was no reason on earth why she shouldn't continue her travels. And so the White House ceremony and the other scheduled events were set—at least for the time being—to go ahead as planned. But for all the meticulous preparation there remained another hitch related to a miscommunication between Marie and Meloney. When shown the deed to the radium, Marie was stunned to read that, according to the paperwork, the gram was to be given solely to her but nothing had been written about what would become of the radium should something happen to her. So long as Marie was alive, everything was fine and the radium would be used only for scientific purposes, as was her intention. But it wasn't clear where the radium would go or what it would be used for if Marie died.

To Meloney's surprise, Marie demanded that the document be drawn up again that very night, so that it was clear the radium was to be an outright gift to the laboratory for scientific uses only. A lawyer, Marie insisted, must be called in to make the changes at once. When Meloney hopefully suggested that, due to the lateness of the hour, the changes might be put off until after the White House ceremony, Marie stood her ground with a youthful resilience that hadn't been evident in the fragile woman who had been sitting with Meloney just a few minutes before. No matter what it took, she asserted, the document had to be revised that same night, and then translated into French so she could make sure there were no further ambiguities. The only question that wasn't answered that evening—probably much to Marie's annoyance—was related to the more than $50,000 in surplus remaining in the fund after a shrewd Meloney managed to negotiate a lower-than-expected price on the radium just purchased. The money was temporarily being held by the Equitable Trust Company, where it would remain for another two years or so before some kind of resolution was reached to every party's satisfaction.

After the deed was rewritten to make clear that Marie had complete control of the radium's present and future use, the ceremony itself went off without a hitch, with Harding paying tribute to the "noble creature, the devoted wife and loving mother who, aside from her crushing toil, had fulfilled all the duties of womanhood." Marie, wearing the same black lace dress she had worn to her two Nobel Prize ceremonies, thanked Harding for honoring her "as no woman has ever been honored in America before." After the presentation, she and the others posed outdoors on the steps before an army of photographers under bright sunshine. Marie wrote later that it was "a radiant day in May" and, from all photographic evidence, Marie herself looked—for once—radiantly happy as she held onto Harding's arm. It was obvious that actually making it to this stage—the moment when she could hold the real deed in her own hands—had done her a world of good. Even so, Mrs. Harding took the liberty at one point of slipping Marie off to a chair at the side so that, briefly, she was able to take a break from the long receiving line that had formed in the Blue Room. (Filling in for her, as always, were the dutiful Irene and the openly sociable Eve, who cheerfully spoke to anyone Marie had missed, in either English, French, or Polish.) In the next day's *Washington Post*, coverage of the event was splashed across the front page in an article that quoted Harding as having said, "The zeal, ambition, and unswerving purpose of a lofty career could not bar you from splendidly doing all the plain but worthy tasks which fall to every woman's lot."

Behind all the pomp and circumstance of that day's celebration was the delicate matter of caring for and transporting Marie's radium. Marie couldn't shake a nagging fear that such a miniscule amount as one gram could be lost in an instant, which is exactly what had happened several years before to the French physicist Henri Becquerel, who accidentally dropped an infinitesimal portion of radium into the crease of his clothing. Like a madman, the scientist had searched and searched, but could never recover the trifling amount, which represented a sizeable fraction of the whole supply of radium in France at the time.

In Marie's case, it was wisely decided that her gram would be split into a dozen different portions, each of which would be placed in a separate hermetically sealed glass vial, all of which would be secured in the lead casket, with walls at least two inches thick. The radium appeared as a grayish-silvery powder, with most of its radiations and emanations absorbed by the glass tubes. The casket, which today resides in the Curie Museum in Paris, weighed about 130 pounds—or nearly 60,000 times the weight of the radium inside it.

Although Marie's visit to the White House was covered by various newspapers, almost nothing was written about any contacts she may have had with the Bureau of Standards, now called the National Institute of Standards and Technology. However, the Bureau was instrumental in measuring and certifying—and then preparing for shipment—Marie's gram of radium. She must have visited the bureau to thank workers for their help. As Marie later wrote, "I have been with special interest to the Bureau of Standards, a very important national institution in Washington for scientific measurements and for study connected with them. The tubes of radium presented to me were at the Bureau, whose officials had kindly offered to make measurements, and to take care of the packing and delivery to the ship."

Before leaving the capital, Marie had been scheduled to dedicate the new laboratory of the US Bureau of Mines. But, after leaving the White House, Marie told Meloney she wasn't feeling well. It was therefore decided at the last minute that she would make the visit but would not be required to trek all the way down to the engine rooms, where the inauguration had originally been scheduled to take place. Officials quickly improvised, telling Marie that, instead, she could stand close to the entrance. Indeed, all that would be required of her was to press a switch designed to turn on all the motors at once. But when the speaker announced in a booming voice, "Now Madame Curie will start the machines in this laboratory," there was complete silence. A few seconds went by and still nothing happened. People in the crowd found themselves squirming uncomfortably. A few even began politely gesturing at Marie with the hope of attracting her attention. But it didn't work. It seems that, despite her exhaustion, Marie had suddenly become engrossed in an examination of a fine specimen of carnotite that had been presented to her as a gift upon her arrival. She turned it over and over in her hands, studying it intently from all angles. Finally, after a few more pokes, an embarrassed Marie quickly pushed the button and the lab's motors roared into action.

Marie later wrote that the programs arranged for her in America were, generally speaking, long and intimidating ones. In order to take pleasure in those parts of the trip more tailored to her tastes, brief as they were, she had been forced to abide by the formalities of many arduous affairs. But no matter how excruciating they were, she always seemed to capture back some of her lost energy when presented with some thing or some place of specific scientific interest—even a single piece of carnotite.

At no time was this more true than on May 26 and 27 when Marie was treated to a tour of Standard Chemical Company in Canonsburg, Pennsylvania, the company that had won the bid to produce her gram of radium. The company had even agreed to produce it for a reduced price of $100,000 in her honor. The company's property was one of only three sites Marie had specifically requested to see. (The others were the Grand Canyon and Niagara Falls.) For Marie, this visit was the best medicine possible, with talk of the kind she thrived on.

The company's history was an interesting one. When their sister had become ill with cancer, the company's founders, Joseph and James Flannery, had traveled to Europe to seek radium for her. After finding out that the supply there was limited, they set up the company in 1911 so they could produce the element themselves. Starting from a radium refining mill on eleven acres just southwest of Pittsburgh, the Standard

Chemical Company had grown into the first successful large-scale commercial producer of radium in the world.

Marie enjoyed listening to the workers, who shared their own thoughts and details about the enormity of the task of extracting radium, a challenge she was all too familiar with, but not on such a grand scale. According to a brief history of Standard Chemical written by certified health physicist Joel O. Lubenau, hundreds of men began the process by digging in a desolate part of southwestern Colorado. He wrote that ore was separated from waste rock by hand at mines, then sacked and hauled by burros to a mechanical concentrator near present-day Uravan, Colorado. The concentrated ore then had to be re-sacked and transported by horse-drawn wagons to Placerville, Colorado to be loaded on trains that were taken to Salida, Colorado. That led to a final 2,500-mile train journey to the radium-extraction plant at Canonsburg, south of Pittsburgh. And after all that, the process still wasn't over. In Canonsburg, even more men used acids, coal, and water to boil and filter the material into a few hundred pounds, which was then shipped under guard to the research laboratories of the Standard Chemical Company in Pittsburgh. All in all, the production of only a few radium crystals from the original five hundred tons of carnotite devoured an almost unimaginable number of hours of labor expended over many months.

It is no surprise, then, that such a minute amount of this precious element would wear a price tag in the early 1920s of $100,000 or more—a value at the time greater than the weight-based cost of the Hope diamond. All in all, it took a total of 500 tons of ore, 500 tons of chemicals, 1,000 tons of coal, 10,000 tons of purified and distilled water, not to mention labor equivalent to that of 350 men for a month, to produce just one single gram of radium. And then the final measurements of the amounts of radium extracted couldn't even be completed until at least four weeks after it had been sealed up in glass tubes, as it takes thirty days for the activity of the sealed radium to hit its maximum potential. Overall, from the time the ore was mined until the time the radium was measured by the Bureau of Standards, more than six months would have gone by.

Marie was riveted. In the end, she spent a full three hours on her feet, touring the facility. The only disappointment was that the man who had helped create the company, Joseph Flannery, was not able to meet Marie. Although he had once told a newspaper that radium would cure "insanity, tuberculosis, rheumatism, and anemia, and a lot of cancers," he himself died of radiation sickness two years before Marie's arrival.

It is astonishing that, as the years went on, the process of purifying radium has stayed essentially the same as it was during Marie's era. However, there has been very little production of radium since the early 1960s. Although the total output of radium since its discovery has been no more than a couple of pounds, much of this is still in existence and is expected to satisfy future demands.

Following her visit to the Standard Chemical Company, the pace of Marie's sightseeing and scientific occasions hardly slackened despite her illnesses. She went on to the University of Pittsburgh, where she was awarded another honorary degree. She wore a gown but, again, refused to don the traditional cap. This had nothing to do with ego or vanity. She simply hated fussing over such trivialities as what one was or was not supposed to wear. The next avalanche of activities included even more honors and events. She received honorary degrees in Philadelphia from the University of Pennsylvania and the Women's Medical College of Pennsylvania. She toured still more laboratories and sat in, as a new member, on a meeting of the American Philosophical Society.

Finally, though, Marie was forced to come to a halt. Worn to the bone, both she and Meloney nearly collapsed one afternoon at the journalist's Manhattan home. High blood pressure, a kidney infection, and general fatigue had gotten the best of Marie while Meloney was suffering the effects of a possibly malignant tumor she had recently learned about—news she had kept hidden from Marie and everyone else. An announcement was put out to the press that Marie had grown too weak and too sick to complete all of her remaining obligations. Under a doctor's orders, Marie would be forced to cancel visits to several cities in the West, including Los Angeles and Pasadena, where all sorts of grand receptions had been planned in her honor for months. Journalists who had been tracking her every move chastised their own country for making Marie "pay with her own flesh for our gift, for the mere satisfaction of our pride." Putting it best was one succinct headline that proclaimed in gigantic letters: "Too Much Hospitality."

Perhaps even Meloney was surprised by the hoopla surrounding their every action. One illustration of just how hard it was for Marie to circulate without creating a major fuss came on May 28, when she attempted to pay a surprise visit to the Memorial Hospital on Central Park West, a place where only cancer patients were received and where $400,000 worth of radium was used in treating them. The scientist's surreptitious plan didn't work. Somehow, word of her coming preceded her and when she arrived, there was a small reception committee on hand to greet her formally. As usual, the preliminaries to what was supposed to be a quick visit around the facility included yet another long round of handshaking and introductions which—as the *New York Times* noticed—"are beginning to bore the French scientist." Even so, Marie was happy to have made the stop. The hospital's four grams of radium represented the most prodigious supply of the precious element gathered in any one place in the world.

According to newspaper reports, Marie's friend Dr. Francis Wood had called the Memorial Hospital only a few hours before she was to pop by in an effort to avoid any dramatics. But hospital authorities weren't having it.

From this point on, as the tour moved ahead on a limited scale, Meloney and others employed every ruse they could to conserve Marie's strength so that—somehow—the rest of the trip could be salvaged and at least parts of the itinerary could go forward as planned. Instead of emerging from trains on the main platform, where frantic crowds assembled, Marie was escorted through less conspicuous exits. But these maneuverings didn't always work. Once, when she was due to arrive at Buffalo, Meloney actually commanded the train conductor to stop at the station before, at Niagara Falls, so that her friend might visit the famous attraction in peace. This time, though, the reception committee in Buffalo got wind of the move. Fans piled into cars and raced by caravan toward Niagara Falls where they crowded around her yet again.

As the days wore on, Irene and Eve were called on more and more often to stand in for their mother, a job they never failed to perform skillfully. Irene, of course, impressed everyone with her scientific knowledge while the charming and dazzling Eve instinctively knew how to captivate an audience. Often, Irene would wear Marie's gown—and even her cap—as she graciously accepted honorary degrees on her mother's behalf. Irene had shown herself to be a brilliant orator. Unfortunately, by now even the normally nonstop Meloney was too exhausted to accompany Marie on at least a few of her upcoming visits, and an American friend named Harriet Eager was called on to serve as the trio's escort on several occasions.

Largely for her daughters' sake, and at Meloney's urging, Marie agreed that a visit to the Grand Canyon would remain on the schedule. For the girls, both lovers of the outdoors, this was probably the best part of the whole trip. Naturally athletic, Irene and Eve both rode horses around the rim of the canyon before giving in to their sense of adventure by taking mules all the way down to the bottom. A surprisingly uninterested Marie, who usually loved the fresh air, uncharacteristically chose to do some shopping instead, buying herself one of the very few pieces of jewelry she ever owned—a turquoise and silver Indian necklace. But the crowds continued to unnerve Marie. At Santa Fe, where they were due to transfer from a private train compartment to a public car, Harriet Eager found Marie sitting all alone, trembling, with her face buried in her hands. In a quiet voice, she told Eager she simply couldn't go on the public car where she feared being stared at, again, like a wild animal. The reason she was so panicked by the crowds there is not precisely clear although she had worked most of her life in laboratories, either on her own or alongside Pierre, and certainly wasn't used to being around a lot of people at once. After a while, with Eager's encouragement, some of Marie's confidence returned, and she managed to steel herself for yet another round of small talk with strangers aboard the train.

The Curies headed back east by way of Chicago, described by the girls as the liveliest of communities. Eve wrote that, "Chicago surpassed all the other [cities] in fervor. [It was] in the Polish quarter of Chicago for a public entirely composed of Poles [where] men and women in tears tried to kiss Marie's hands or to touch her dress."

During these final weeks of the trip, both Irene and Eve made memories that would stay with them the rest of their lives. When they weren't representing their famous mother, the two sisters were playing tennis, going boating, and, according to Eve, enjoying excursions such as "an elegant week end on Long Island, an hour's swimming in Lake Michigan, a few evenings at the theater, and a night of wild delight on Coney Island." They were especially enchanted by the three days on the Santa Fe line train traveling across the "sands of Texas" and vowed never to forget the charming hotel at the Grand Canyon, "an islet of comfort on the edge of that extraordinary fault in the earth's crust."

By June 17, Marie was again the very image of a woman beaten down and so, for the second time, the journey was interrupted. Her blood pressure, which was now terribly low, alarmed the doctors. They ordered her to stay off her feet. From then on, following days of doing almost nothing but chatting with Meloney, she was able to visit only a few more universities and laboratories across the Northeast before calling it quits.

In any case, it was now time to go back to France. Despite Marie's physical state, the American tour couldn't have been more successful from a financial point of view. According to a letter from Meloney to Yale University, Marie had come away not only with her gram of radium and $22,000 worth of ores, she had also amassed a large assortment of expensive equipment and nearly $7,000 in cash awards from different groups. In addition, more than $50,000 in surplus money remained in the Marie Curie Radium Fund in a New York bank and its interest would eventually provide Marie with another source of personal income for the rest of her life.

Always for Marie, though, it was that half teaspoon of material that would soon be locked away in the purser's safe on the ship that was of paramount importance. In addition to getting back on track financially, Marie had also managed, on a separate level, to tunnel through her fears and come out on the other side. Langevin-Joliot said that those weeks in America forever shattered any notions her grandmother may have harbored about somehow being able to live out the rest of her life in the shadows. They also showed that she could sit under the spotlight—and survive. From then on, Marie talked more openly with her daughters and associates about everything going on in her world. She met with more people. She socialized. "She was out in front more," Langevin-Joliot added.

The trip's only real disappointment was that the Manhattan eye specialist Meloney had brought in to help Marie wasn't able to do so because her cataracts were much worse than originally thought. As Marie said goodbye to Meloney, she obviously feared she was going blind. "Let me look at you one more time, my dear, dear friend," Marie said. "This may be the last time I will ever see you." Marie thanked Meloney for all she'd done, and both women wept as they embraced farewell. Irene and Eve, too, were sad to say goodbye to their companion—and to what had been a spectacular adventure. On June 28, the Curies embarked in New York on the same luxurious liner that had brought them to the United States less than two months before. Walking into her cabin, Marie was bid a final glorious sendoff with a mountain of telegrams and bouquets. From the ship, Marie wrote another personal message of thanks to Meloney, saying that "you helped me in the most unselfish way in the work I love and I wish that I could be helpful to you in something you would want very much to do."

On the heels of such a successful trip, almost entirely orchestrated by Meloney, the Curies would call on the journalist again and again in the months and years ahead whenever they needed help from her part of the world. Even a young Eve poured out her requests on paper, asking Meloney just a few months later, in November 1921, if she wouldn't mind helping arrange a concert tour in the United States for her friend and occasional piano teacher, the noted Russian-born French pianist Alexander Brailowsky. Eve wrote that she was quite ashamed to be bothering Meloney with such an issue, but that "mother says you are so very good that you will pardon me my indiscretion." Eve asked Meloney to use her personal influence in order to help organize a series of concerts in the United States as well as positive media coverage. As always, Meloney agreed to do what she could and ultimately Brailowsky made his US debut at Aeolian Hall—at the time a concert hall near Times Square where the world's leading musical figures played—in November 1924.

Later on, still another letter has an even bolder Eve asking Meloney for help breaking into the American journalism scene. The budding writer suggests she could write about fashion and theater, noting that she also could do interviews with famous people. "I would be extremely grateful to you if you could just tell me if you think it possible for me to find that business and how I could find it," Eve wrote.

Marie, too, wrote Meloney many times with both minor and major requests. Upon her return to France, Marie wrote Meloney about all the letters she'd received from schools across America, asking her for autographed photographs and messages. What was she to do with them? In this case, it's not clear what Meloney advised. Marie also asked Meloney—quite often—for advice related to Marie's biography of Pierre, conveying an implicit trust in her friend's opinion. In September 1921, Marie told Meloney not to go forward with publication "if you are in doubt." Marie said that the biography could no doubt be kept and shared without ever being formally published, emphasizing that she would not like to have anything published unless it gave Meloney pleasure. As early as 1921, Marie began hinting in her correspondence to Meloney that the journalist's help might be called on yet again, in a big way, to help fund a Radium Institute in Warsaw. Indeed, this challenge would lead to yet another trip to America down the road. But first the Radium Institute in Paris would have to be looked after. In November 1923, Meloney sent Marie a cable asking for a list of emergency equipment needed to complete the institute's laboratory. Marie responded that she could still use all sorts of material. She invited Meloney and other Americans to choose to provide the items off a list prepared by Marie, adding that she "would be very grateful."

From her writings, it appears that this flood of requests never seemed to bother Meloney. She happily used her influence to benefit the Curies many times over the years to come. She seemed to adore all three of the Curie women and always relished any opportunity to assist them.

Following days of fairly calm seas, the *Olympic* finally arrived back at Cherbourg, France, and Marie and her daughters went on to the Gare Saint-Lazare in Paris by train, arriving on July 2, 1921. Rather than hordes of cheering fans, just a handful of men were there to greet them:

among them were two reporters, their scientist friend Jean Perrin, and a young lab assistant. Marie and her small entourage couldn't even immediately hail a cab. Apparently all the drivers were distracted by accounts of a world-championship fight between France's Georges Carpentier and America's Jack Dempsey that were being played over loudspeakers in the streets. A reporter asked Marie's opinion on the match and Marie responded, "I regret that I have no opinion on the subject." At last, Marie and her daughters were home.

CHAPTER 6

New and Improved

Like New York City, and all great cities, Paris is a composite of many worlds. It was a place known by the late 1800s as home to the world's most celebrated medical schools, attracting thousands of students from around the world. Marie Curie had come to Paris before the turn of the century as a young woman in her twenties to see and to learn—and she had grown to think of the city as her own. From the moment she laid eyes on it, she adored the bookstores, the monuments, the architecture, the straight avenues and, most of all, the famous Sorbonne, described even centuries before as "an abridgement of the universe." At the time, she thought the adventure of Paris fit for a fairy tale.

The trip to America skewed her vision. She still felt a great affinity for Paris even though many in France had labeled her a traitor only a decade before. But the touring of so many quality medical centers and highly funded laboratories and universities had left a bitter taste in her mouth. Not so long ago, France had set the international benchmark when it came to medicine and learning. Not anymore. Now, obsolescence loomed.

By 1921 the emerging eminence of hospitals in America was making it the only place to be for the world's talent pool of ambitious researchers. The country's universities were bursting with creative energy and benefiting from a flood of new investment. Most bothersome to Marie was that, at the time, France wasn't home to a single hospital devoted to the treatment of cancer patients with radium—Curietherapy. With its rich supply of radium ore, America, on the other hand, housed several.

Would this imbalance be a permanent one? Not if Marie had anything to do with it. But she had no inkling of what was to come. The radium-producing industry that had started in France but had taken off more rapidly in America was about to turn into an unexpected deathtrap for unsuspecting workers.

Initially, though, the journey to America had the effect of an alarm bell. After behaving like a monk in Paris, Marie embraced a new calling, which demanded nothing short of embracing a new way of living. She returned with a new vigor that she poured into her Radium Institute, determined to take what she had learned about the power of American-style public relations and use it as a means to achieve even more than she had previously thought possible.

In America, an event or cause could be made—or broken—by her mere presence. Her star power, she discovered, was an asset she hadn't recognized. Even for such a modest woman, it must have been an intoxicating realization. After 1921, Marie would become more driven by human relationships. She would put up with the burdens of overseas trips, public ceremonies, and social engagements. Instead of relying on a small coterie of like-minded physicists, Marie reached out to a larger circle of people. "She put herself out in front of people more," Hélène Langevin-Joliot noted. She said America taught her grandmother a simple but valuable lesson: if you wanted something, you had to ask for it. "She finally accepted the idea that she had to make things known to people," she said, adding as an aside that she still, to this day, can't get over the fact that those in America celebrated the importance of her grandmother's work before many of those in France.

In general, Marie's granddaughter had a lot to say about the trip that almost didn't happen.

"She was not the same person when she came back here," Langevin-Joliot said. "She finally understood just how important she could be."

Marie returned to France as a fifty-three-year-old woman with the means and the skill to turn her intention of creating a world-class research institution into a reality. Already a retinue of employees—many of them women—was counting on her. Irene was more than ready to be groomed to fill her mother's shoes. Clearly, more discoveries were waiting to be made.

A major priority was the translation of her fame into continued support for the work of the next generation of scientists at her institute. As Meloney described it, Marie wanted to spare other young researchers from the hardships that she and Pierre had endured. Going forward, she would expend enormous energy not only on tedious social exercises, but also on the courting of government officials, always with an eye toward earning fellowships for talented students.

As Langevin-Joliot explained, young people working in labs in the twenty-first century are often forced to help find funding for their own work. "But in the old days the director of the lab [in this case Marie] got the money while the young people just worked and that's the way Marie wanted it," she said.

After returning from America, Marie ruminated publicly for the first time on why she and Pierre had always refused to profit from their discoveries. Perhaps she was feeling the need to reassure herself and others that she wasn't breaching her cherished scientific tenet of disinterestedness by turning to the United States for help. She wrote in the autobiographical notes prepared for Meloney that, from the very start, she and Pierre had never hesitated when asked to share information about how to prepare radium. They never doubted their decision not to take out a patent or derive an advantage from any industrial exploitation. She wrote, "No detail was kept secret, and it is due to the information we gave in our publications that the industry of radium has been rapidly developed. Up to the present time this industry hardly uses any methods except those established by us. The treatment of the minerals and the fractional crystallizations are still performed in the same way, as I did it in my laboratory, even if the material means are increased."

During her commentary, an admission spilled out. Marie acknowledged that she'd sacrificed a "fortune," money that could have been passed on to her daughters and could have even prevented her from needing America's help. In fact, if she had patented the production process, she would have had the financial means to create an outstanding Radium Institute many years earlier. "Yet I still believe that we have done right," she wrote. And this willingness to give to humanity without material benefits, Marie explained, is what had most appealed to the generous nature of Americans. "Our American friends wanted to honor this kind of spirit," she added.

And in the years to come, Americans would honor it over and over. Under Meloney's guidance America had indulged Marie once—and it would do so again.

Although resentment remained an incorrigible relic of the trip, Marie would be eternally grateful for the gram of radium, her precious element, even if it was about to spill many unforeseen sorrows.

Before carving out the future, though, Marie and her daughters traveled together to L'Arcouest for a much-needed respite after so much strenuous travel in America. Then, in September 1921, Marie left for the seaside resort of Cavalaire in Provence while the girls stayed on in Brittany. For a brief period, according to her letters, Marie gave herself permission to do nothing. "I have hardly worked at all," she wrote from the sunny village located on the Mediterranean coast. Once again, Marie missed Irene's birthday as her daughter turned twenty-four on September 12. She wrote Irene that she couldn't think of her "more tenderly" than she did on that day. She told Irene that she had made her life so much sweeter. Irene replied that "when you're not with me I feel that something is missing."

More and more, their relationship was one of mutual respect marked by a special bond that appeared to run deeper than the one between Marie and Eve. In one letter from Marie to Irene, she wrote about the changing weather, noting that "we will have to conciliate our scientific work interests represented by the two of us with that of the musical arts represented by Evette, which is much easier in nice weather than in the rain." Although Marie had hired many music teachers for Eve over the years, and purchased a grand piano for her, it seems she was bothered by the idea of being holed up inside—in bad weather—with Eve while she practiced piano. At the age of nineteen, Eve still had little idea what she wanted to do with the rest of her life, just the opposite of her sister's single-minded pursuit of science.

Upon returning to work in Paris, Marie allowed herself a bit of rare self-indulgence. At the urging of her American friends, she finally installed central heating at her Paris apartment. She also purchased her own piece of property in L'Arcouest. After vacationing there for so many years, she and her daughters would no longer have to rent a place or stay in the homes of friends. L'Arcouest was as picturesque a place as there ever was, a cluster of houses on the rocky shore of Brittany, and it had become a magnet for Sorbonne professors and their families. Marie and her daughters always said that many of their happiest memories were made during their holidays there.

But before long Marie would get back to the task of finishing off her Radium Institute. When it first opened, on July 31, 1914, Marie had only four days to foster it before World War I broke out. Seven years later, all that remained was a shell of what Marie had imagined it would be. The simple inscription—"Institut du Radium, Pavilion Curie"—was still etched into stone at the entrance. And a garden of climbing roses, planted by Marie, still bloomed spectacularly each year. But she needed the finest equipment and the most gifted employees to cap it off. And—at last— she could afford it. Prospective researchers came out of the woodwork. Generous to a fault, if a close friend or respected colleague recommended someone, Marie almost always found a place for them. She had a particular fondness for women and foreigners, especially Poles. Indeed, in the two years after the war, women still made up the lion's share of Marie's staff as so many men had died in combat.

To many she seemed cold. Yet workers at the institute always gave Marie credit for taking a personal interest in their health by forcing them to go outdoors each day so that they could breathe in some fresh air. And many showered her managerial skills with praise. One chemist described the institute as "a unique place of work and meditation, isolated from the world, within which [Marie] had gathered those she accepted as working companions, whether they were scientists of high quality, beginning researchers or modest technical collaborators." Like radium itself, the institute was now Marie's other baby, and she fretted over it whenever she was away. She needn't have worried though. By this time, an accomplished Irene had no qualms about taking charge during her mother's absences.

In matters of experimentation, Marie wrote that the radium from America was being devoted to research on the radioactive transformation of atoms.

"It is a well-known fact that scientists have not been able till now to alter the course of these transformations by any means at their disposal and this leaves us utterly in the dark as to the possible reasons of the transformation," she wrote.

Marie acknowledged that atoms of radium break up from time to time, producing spontaneous radiation, while atoms of lead, gold, and other metals do not exhibit radioactivity.

"But why it is so we could not tell . . . if this could be done, the experiments could throw light on the cause of the atomic change and on the atomic structure," she wrote.

There was still so much she and others didn't know.

Upon her return to Paris, Marie was somewhat taken aback to discover that her warm reception in America had suddenly improved her standing in the country that had—in 1911—nearly booted her out. And the new Marie took full advantage of it.

In February 1922, she was elected to the Academy of Medicine, a move that shattered a 224-year tradition of excluding women. (There were 64 votes cast in her favor out of a possible total of 80. Many ballots were turned in blank, proof that many were still opposed to women being admitted.) But the election had occurred spontaneously without the regular ordeal of a candidacy, a testament to Marie's reputation. The academy's president praised Marie not only for being a great scientist but also for being a great woman and patriot, saying that "Your presence here brings us the moral benefit of your example and the glory of your name. You are the first woman of France to enter an academy but what other woman could have been so worthy?"

Although Marie still didn't care much about accolades, she knew that membership would lead to increased contacts with the medical profession. And she wanted desperately to be viewed favorably by European doctors in order to accomplish two of the goals she and Pierre had worked toward in their last years together: the development of Curietherapy and the establishment of safety standards for researchers. In a letter to her sister

Bronya, Marie had noted that it was possible that radium had nothing to do with health issues, "but it cannot be affirmed with certainty." Because of her dubiety over radium, Marie had been one of the first to require her workers to undergo routine blood checks.

Three months after the election, another important route of influence opened up to Marie when, by unanimous vote, the Council of the League of Nations named her a member of the International Committee on Intellectual Cooperation, an arm of the League of Nations established in 1922 and financed by the French government and—eventually—by many other governments as well.

Again, infused with a new perspective on her place in the world, she accepted. After all, some of the most esteemed scholars had been asked to join, most notably her friend Albert Einstein. (The first American member was Professor George Hale, the astronomer, in 1922.) But Einstein's on-again-off-again membership was draped in controversy—with Marie caught in the middle.

Marie and Paul Langevin, still very good friends, had recently asked Einstein to lecture at the Sorbonne. But the animosity between France and Germany following the war was so potent that several members of the French Physical Society, especially the anti-Semites, had threatened to protest the event. Einstein was born of Jewish parents in Ulm, Germany, in 1879. But he was so opposed to German militarism that he took Swiss nationality in 1901. Marie and Paul opted to go ahead with the lecture at the Sorbonne—and a large crowd attended. But another talk scheduled at the French Academy had to be canceled at the last minute when thirty members vowed to disrupt the event and walk out if Einstein did indeed show up.

The avoidance of just these kinds of conflicts was precisely why the International Committee on Intellectual Cooperation had been set up in the first place. In fact, the committee's overriding purpose was to bring together intellectuals of different nations who had been isolated by the war. (Germany, though, was denied membership in the League of Nations until 1926.) Einstein initially accepted the invitation to join the committee. But he resigned before ever taking part, after nationalist fanatics assassinated his friend Walter Rathenau, Germany's Jewish foreign minister, in June 1922. Einstein, too, received death threats and was warned not to be part of any political groups deemed controversial. Greatly disappointed, Marie implored Einstein to reconsider, writing: "It is precisely because dangerous and prejudicial currents of opinion do exist that it is necessary to fight them. . . . I think that your friend Rathenau, whom I judge to have been an honest man, and whose sad fate I regret, would have encouraged you to make an effort at peaceful intellectual collaboration. Surely you can change your mind." Although he harbored reservations, Einstein eventually succumbed to pressure from Marie and, despite the continued threats, rejoined the committee.

But then, in the fall of 1922, Einstein refused to attend the next meeting of the League of Nations. This time he was angry over the group's failure to respond after French and Belgian troops took control of the Ruhr district in Germany. The push into the Ruhr was an attempt to claim the area's coal and iron after Germany failed to make reparations in the aftermath of the war. But the aggressiveness of the act dismayed Einstein. He told Marie that, although he was sorry to disappoint her, he simply couldn't be part of the group after all. In his mind, the League of Nations was nothing but a tool of power politics giving off only the illusion of objectivity.

Marie stepped in again, writing that she agreed that the organization wasn't perfect but that it had no chance of being so since humans aren't perfect. Yet, she insisted, it could make a difference, noting that "it is the first attempt at an international understanding without which civilization is threatened with disappearing." In a sign of just how much he respected Marie, Einstein once again changed his mind.

As Eve wrote, Marie's membership on the committee marked a pivotal moment in her mother's life. Since she had become a celebrated scientist, hundreds of leagues, charities, and associations had begged Marie to lend her name in support of every cause imaginable. But she had almost never allowed this. Marie hadn't wanted to become a member of anything that might take her away from her work. Just as important had been her desire to maintain neutrality in all political matters. She refused to abdicate her high title as "pure scientist" in order to throw herself into the quicksand of varying opinions. As a result, even those behind the most inoffensive of manifestos couldn't garner her signature. Eve said that becoming a member of the International Committee on Intellectual Cooperation was to be Marie's only real infidelity to scientific research.

Eve described Marie's work on the committee this way: "Finally—paradox of paradoxes—the physicist who had always avoided material profit for herself became the champion of 'scientific properties' for her confreres: she wanted to establish a copyright for scientists, so as to reward the disinterested work which serves as a basis for industrial applications."

Although Marie had never sought financial gain for herself, she wanted to help financially strapped laboratories by obtaining subsidies for pure research—even if she did have to tap into the profits of commerce.

At a meeting of the committee, Marie persuaded fellow members to adopt her suggestion to have "scientific vacation programs in which young scientists, both men and women, from one country, would go to another during the summer." Yet it would be some time before she found a government willing to finance it.

All her life, Marie had been obsessed by a certain assumption: that intellectual gifts were being wasted among the classes that hadn't been blessed by good fortune. She believed that inside every peasant could be hidden a writer, a painter, a musician, or, of course, a scientist. It was up to her, she thought, to develop international scientific scholarships so that these struggling but capable individuals might be able to uncover and cultivate their innate abilities. Often she wrote of her frustration at her lack of progress in ameliorating their plight.

The following year, on December 26, 1923, Marie enjoyed perhaps the most personally touching tribute of her life when a special event was thrown in the Sorbonne's amphitheater to celebrate the twenty-fifth anniversary of the discovery of radium. Students, teachers, friends, loved ones, politicians, and fellow scientists turned out in droves. The president of France, Alexandre Millerand, presided over the festivities during which her close friend, scientist André Debierne, read before the audience the 1898 report announcing the discovery of radium titled "On a New and Strongly Radioactive Substance Contained in Pitchblende" by Pierre Curie, Madame Curie, and G. Bémont, Looking on with enormous pride, and with eyes brimming with tears, were two elderly women alongside a silver-haired man. Marie's sisters Helena and Bronya and her brother Jozef had come all the way from Warsaw to share Marie's moment in the spotlight in France. To celebrate the occasion, the French government passed through parliament, by unanimous vote, a law granting Marie an annual pension of 40,000 francs as a "national recompense." Upon Marie's death, Irene and Eve were to inherit the money.

As had happened many times before, Marie was again perhaps the only one unable to fully enjoy the proceedings. Although she had undergone a cataract operation only a few months before under an assumed name, the surgery hadn't worked and her eyesight continued to deteriorate. For weeks, Eve played dutiful nursemaid as Irene was off camping in the countryside with friends. Later, Eve conceded that the handling of so many radioactive substances had nearly made her mother blind.

Initially, Marie and others were so fascinated by the mysterious properties of radium that they failed to fully grasp—or accept—the dangers associated with it. But then grim descriptions of troubling health conditions began pouring out of the United States, where the radium industry was booming.

Most disturbing was the saga unfolding at US corporations that had employed workers to paint luminous numbers on watch faces using a powdery radium mixture that glowed in the dark. These watches, originally designed for World War I soldiers, had become all the rage by the 1920s. Big corporations were only too happy to cash in, and they paid a decent wage to those willing to sit for hours on end in large, dusty rooms at long tables stacked with dials waiting to be painted. Lured by the money, a multitude of mostly young women applied. They had no idea that they were stepping into one of the most dangerous work environments on the planet.

To paint the dials, the women used brushes made from camel hairs, brushes that would lose their shape after only a few strokes. As a result, they were instructed to mold them into fine points—by using their mouths. It never dawned on them that they were ingesting traces of radium with every lick.

One of the workers was a woman named Grace Fryer, dubbed "Pretty" Grace Fryer by the newspapers, who began working for the US Radium

Corporation in the sleepy New York City suburb of Orange, New Jersey, in 1917 along with seventy other women. As had become routine, she tapered the brush hairs with her mouth.

One day, Fryer had to blow her nose. She later noticed that her handkerchief glowed in the dark. For a few minutes, Fryer might have wondered what was going on. Even so, no one seriously imagined they were being poisoned. After all, most of the women actually liked the material's luminescence, painting their lips and fingernails with it in order to enchant their boyfriends. Fryer was no different. She only quit the US Radium Corporation when she was offered a better-paying position as a bank teller.

By this time, though, dangerous levels of radium had been deposited in her bones. Within two years, Fryer's teeth began falling out. Then her jaw became unbearably painful and her eyes grew clouded. She consulted a doctor, who discovered serious bone decay in her mouth. But, even after a thorough examination, he could find no reason why this decay should be occurring. Yet the pain continued and Fryer became weaker. Finally, in 1925, a second opinion surmised that her health issues might have something to do with her old job at the radium corporation.

By this time, the situation had become more harrowing than anyone could have imagined, a story Marie kept up with from France.

On May 29, 1925, Dr. F. L. Hoffman, a Prudential Insurance statistician, told a meeting of the American Medical Association that five women had died and ten more had been struck by a painful swelling and porosity of the upper and lower jaws—or "radium jaw." He suggested that this new occupational hazard could easily be avoided by simply asking workers not to lick the brushes. But the directors of the US Radium Corporation weren't buying it. They refused to admit any link between radium and the deaths, citing Marie Curie as an example in their defense. "No one has worked longer or with greater amounts of radium than has Mme Curie," the owner said. "For over 25 years she has toiled increasingly in her laboratory and today is not only alive but reported recently to be in excellent health." But in the face of mounting evidence to the contrary, the company was eventually forced to change its tune, finally admitting that radium could have had something to do with the health problems suffered by the painters. The company ordered that all brushes be sterilized before use—and that employees not lick them.

Fryer's plight, and the cases of the other women, might never have become public if not for the efforts of the National Consumers League and the journalist Walter Lippmann, an editor at the *New York World*. For its part, the National Consumers League had fought for better working hours and pay for women since 1899. Lippmann, meanwhile, was a crusading reporter representing an influential New York newspaper. Both helped bring news of the women's condition to mainstream America and to scientists around the world.

From France, Marie kept tabs on the case and told reporters that she had never heard of anything like it "not even in wartime when countless factories were employed in work dealing with radium." But she pointed out that French radium workers utilized small sticks with cotton wadding—not paintbrushes. She reminded the public again that she wasn't a doctor but said she'd be "only too happy to give any aid that I could." When a New Jersey journalist asked Marie if she'd discovered anything that might help, noting that a dozen women had already died, Marie said there was no way to get rid of the radium once it had entered the body.

When Pierre was alive, he and Marie had always accepted the risks associated with radium as the price a serious researcher had to pay. But in time Marie began to reconsider her tough approach. She had no other choice. The number of people getting sick was just too substantial to ignore. And so, by the mid-1920s, Marie was not only requiring her workers to undergo regular blood tests but also to employ masks and tongs. She doesn't seem to have taken any extra precautions herself, however.

In Europe, where the war had resulted in millions being X-rayed, many people who had been exposed to radiation were also starting to suffer from an array of ailments. Despite the benefits afforded by X-rays, an army of doctors, technicians, and scientific workers now seemed to be in danger. One scientist acquaintance of Marie's died not long after spilling a highly radioactive substance on herself. Within four days of each other in 1925, two other acquaintances of Marie's, engineers named Marcel Demalander and Maurice Demenitroux, died of leukemia and severe anemia just months after handling radioactive substances. Demalander was thirty-five while Demenitroux was forty-one. Immediately a charitable Marie organized a fund to raise money for their widows.

Then, in 1927, while Marie was away in L'Arcouest, Irene wrote her about a Polish chemist named Sonia Cotelle who was "in very bad health . . . she has stomach troubles, an extremely rapid loss of hair, etc." She had been working with polonium. Not overly concerned, Irene also wrote that, "given that I have worked a lot on that without being made ill, I think more that she must have swallowed some polonium. . . . What's more, her present ill health perhaps has no connection with that but she is very uneasy, which is understandable." Although Irene seemed hesitant to outright associate Cotelle's troubles with radioactivity, Marie by this time thought the news worrisome enough to release her own warning to researcher friends in America about the possible dangers affiliated with the handling of radioactive materials. But at the same time she also added her reassurances that she knew of no "grave accidents due to radium or to mesothorium among the personnel of other factories . . . nor among the personnel of my institute."

Marie's own bouts of illness were growing more regular: the anemia, weakness, overall fatigue, and drops in blood pressure, not to mention the problem with her eyesight, were by this time an accepted part of her daily world. She also began tolerating an almost constant low-grade fever. All were undoubtedly due to her lengthy exposure to radioactivity. But she was obsessive about keeping her ailments private, always going by an assumed name when seeking treatment. Of course, Marie hated publicity. But she also hated the idea that illness might one day force her to retire. She confessed to her sister in 1927 that, even if she was still writing scientific books, she wasn't sure she could live without the laboratory.

While more and more people were starting to view radium with suspicion, there were still those who were touting it as the be-all and end-all answer to seemingly every human ailment. Marie's great friend Meloney, who wrote about the plight of the watch painters in 1925, underwent experimental radium treatment for a possibly malignant tumor and lived another twenty years. Marie's other American friend, the dancer Loie Fuller, also successfully underwent radium therapy when she was diagnosed with breast cancer. In addition, the mass market was peddling everything from a "Curie Hair Tonic" that was supposed to stop hair loss to a "Crème Activa" that was supposed to promote eternal youth. Time and again, radium's benefits were inflated out of all proportion.

Even so, an explosion of newspaper articles portraying the dangers of radioactivity spurred governments into action. In 1921, the British X-ray and Radiation Protection Committee was organized while, in 1928, an International Radiation Protection Committee was formed. The United States Advisory Committee on X-ray and Radium Protection was the first to set safe X-ray exposure limits. In other countries, too, organized efforts took shape to establish definitive standards for X-ray shielding. In 1925, at the request of the French Academy of Medicine, Marie joined one such commission to improve safety measures. In the end, measures were adopted that recommended enclosing radioactive material in thick, heavy metal and requiring researchers to work behind lead screens and to use tongs rather than fingers to maneuver radioactive material. Blood was to be routinely tested for abnormalities.

Yet any safeguards imposed by these groups would come too late for many who had already worked with radioactive materials, including Marie—and her precious daughter Irene.

CHAPTER 7

Another Dynamic Duo

On December 16, 1924, a dashing twenty-four-year-old military officer and aspiring scientist named Frédéric Joliot walked into Marie Curie's office to apply for a job as a junior assistant. As a boy conducting messy experiments in the kitchen sink, he had been so enamored with both Marie and Pierre that he had taped a photograph of the couple to his bedroom wall. Now the slim grown man with a shock of thick black hair was hoping to become his hero's indispensible deputy.

Later he recalled how nervous he'd been during his initial meeting with the woman he'd put on a pedestal; she was petite with gray hair and kind eyes—but intimidating nonetheless.

At one point during the interview, she tersely interrupted him as he was carefully reciting his answers to questions about his background, according to Denis Brian's biography "The Curies."

"What! You don't have your [baccalaureate] degree?" She said that if ever hoped to teach, he's need to obtain more advanced educational qualifications in addition to a teaching license.

Taken aback, he simply nodded his head in agreement.

"Can you start work tomorrow?" she asked bluntly.

"I have three weeks of [military] service to complete," he answered.

"I will write to your colonel," she said before abruptly putting her head down and going back to her reading.

But Joliot's interview hardly mattered anyway. He came highly recommended by Paul Langevin. Although the love affair between them was long over, Marie spoke regularly with Paul and if he asked for a favor then, of course, she obliged.

And so Marie asked Frédéric to start the very next morning as a technical assistant. The move ultimately led to the creation of yet another dynamic duo in science.

Born in 1900, Frédéric was the sixth and last child of a moderately prosperous Parisian family that was, like the Curies, not religious. His father, Henri Joliot, had fought for the Paris Commune, the worker's revolution of 1871, and had been exiled from France until an amnesty brought him back in 1880. As a young boy, Frédéric was sent to boarding school at the Lycée Lakanal just south of Paris, where he would set himself apart more in sports than in academics. But after his father died, the family suffered a reversal of fortune, and Frédéric was forced to attend the free vocational school where Pierre had once taught. There was no way the family could afford to send him to either of the city's two best schools. Nevertheless, he made the most of his situation, studying hard and eventually graduating first in his class. Because the army was short on men, Frédéric was mobilized when he turned eighteen, two years earlier than expected. As he neared completion of his military duties, Frédéric applied for and won a research grant and, on the recommendation of Langevin, wound up in Marie's office at the Radium Institute.

Langevin, who served as director of studies at the vocational school Frédéric attended, had been the first to take notice of his potential and especially his talent at experimentation. Langevin often invited Frédéric to his home, where the two sat in front of the fireplace for hours discussing science, politics, and philosophy. Langevin steered the young man toward scientific research and also toward a pacifist and socially conscious humanism that ultimately led him to socialism. After his first three terms of school, Frédéric had chosen physics as his major, at Langevin's urging. As his skills matured—honed by a curriculum largely devoted to laboratory work—he displayed exceptional manual dexterity and an unusual originality in his thinking,

Frédéric's closeness with the highly revered Langevin couldn't have been any more fortuitous. Since, like Pierre, Frédéric had not attended the country's elite schools, he might very well have written off any chance at a scientific career. But thanks to Langevin he was able to worm his way in. The married Langevin, who had enjoyed a love affair with a two-time Noble laureate, didn't mince words when he told the student that science was a career with the ability to dissolve all hopes of a normal private life. He also warned Frédéric that France's scientific community was very much a professional and even a social clique that was hard to break into. But he hailed science as a noble calling. To assuage any fears Frédéric might have harbored, Langevin said he had confidence that he could distinguish himself by making a great discovery. Langevin opened a lot of doors for Frédéric and was destined to be one of the three main influences in his life, the other two being Marie and Irene Curie.

Despite his enthusiasm for the job, Frédéric's first two weeks at the Radium Institute could have gone better.

Just after Frédéric started, Marie's close friend and second in command, André Debierne, warned Frédéric that since the world already knew so much about radioactivity there was little more to be done in the way of new research. Others were also less than encouraging. Indeed, Frédéric initially found the institute's workers and regimen so off-putting he wondered whether he might be in over his head.

But then an auspicious introduction turned things around. On the very first day of 1925, Marie placed Frédéric in the charge of her very aloof daughter Irene. There is no indication that Irene had enjoyed any previous love affairs—although she was good friends with Einstein's son and enjoyed close relationships with Debierne and her cousin Maurice Curie—or that she showed any romantic inclinations. And so she probably didn't give a second glance to her dashing new protégé, who was three years her junior. Irene only had one thing on her mind when she was at the institute: work. And so, without knowing it, she gave Frédéric the worst assignment possible for someone with his limited experience: to study the chemical aspects of polonium and other radioactive elements, an area about which he knew almost nothing. At first, he must have felt even more demoralized. But then he decided he must master the task if he was to stay at the institute—and master it he did, even surpassing Irene's expectations. Not only did he follow her instructions to the letter, but he also went beyond them by discovering new ways to modify various apparatus so that the experiments could be conducted more successfully.

Frédéric probably wasn't sure what to make of Irene. Like everyone else, he most likely tagged her as cold and distant. She had always been exceedingly blunt and gruff. At the institute she often was so preoccupied that she failed to utter even a quick hello to colleagues when she arrived each morning. As a result, most considered her to be unfriendly; she was too remote and, some thought, too similar to her tough mother in temperament. But no matter how they felt about her personally, no one could discount her incredible intelligence and talent. She earned her position through merit—although no doubt Marie had always paid special attention to her daughter's career.

Working with Fernand Holweck, chief of staff at the institute, Irene performed several experiments on radium, which had led to the writing of her first scientific paper in 1921. It wasn't long before Irene had published several articles on the high-speed particles known as alpha rays including one on the distributions of rays that was promptly passed on to Albert Einstein by an extremely proud Marie.

In July 1924, only a few months before Frédéric was hired, Irene wrote Marie in Geneva to tell her how pleased she was by her studies of the magnetic deviation of alpha rays, noting that her findings were so precise that they were superior to those achieved with the Geiger counter. (She also noted that she'd been practicing her swimming crawl.)

Marie wrote back that she was "delighted" and urged her to send on her findings so that she could share them with Einstein. Apparently, Eve had accompanied Marie to Geneva, where she was attending a gathering of physicists. "Very luckily, Eve is not bored. . . . [S]he seems very happy to see a lot of Mr. Einstein, who is very kind to her," Marie wrote.

Beginning in 1925, Irene began making her own headlines. On March 30, three months after meeting Frédéric, Irene publicly presented a defense of her doctoral thesis on the emission of alpha rays from polonium—the laboriously titled "Research on the Alpha Rays Emitted by Polonium: Oscillations of the Trajectory, Initial Velocity and Ionizing

Effects." She must have asked Frédéric to read it as she told him once that she hoped it would not be as boring to read as it had been to write.

It's hardly surprising that more than one thousand people turned up at the Sorbonne's amphitheater to hear what the daughter of the famous scientist had come up with. Indeed, reporters at papers from around the world, including the *New York Times,* covered the event. Even before she said a word, more than one had predicted that Irene would be the next female Nobel Prize winner.

Donning the same sort of simple black dress her mother had worn for a similar event more than twenty years earlier, and wearing unstylishly short hair, Irene had calmly collected notes from her apartment that morning and coolly strolled to the nearby university where—despite the huge crowd—she delivered her presentation in a confident and booming voice. She explained that, in an attempt to learn more about the nucleus of atoms, she had bombarded it with polonium. She had become intrigued at the movement of alpha particles as they were given off by polonium during natural radioactive decay. Polonium, of course, was the radioactive element her mother had discovered, which Irene acknowledged by dedicating her thesis to "Mme. Curie by her daughter and her pupil."

No doubt many in the audience had little idea what Irene was talking about, but when she finished, those in attendance applauded as if they did. The crowd's enthusiasm was shared by the professors who were there to decide whether or not she deserved a doctorate.

As Eve summed up the day, Irene went "calmly off to the Sorbonne, came back confident of having passed, and waited without any great emotion for a result that was guaranteed in advance." In an interesting aside, Irene's doctoral professor at the Sorbonne was Paul Langevin, who had—many years earlier—been supervised by Irene's father, Pierre.

Afterwards, to celebrate Irene's achievement, Marie hosted an elegant tea and champagne party in the garden outside the Radium Institute to which the entire staff, surely including Frédéric, was invited. Cookies were laid out on developing trays from a photographic darkroom while tea brewed in laboratory flasks was served up in beakers. Langevin arrived just in time for a champagne toast and predicted a future full of success for the newly minted scientist.

Notably, Irene displayed less hostility toward the press than her mother did. When a young female reporter asked her a few questions following the event, Irene didn't hesitate to respond. When asked about women and science, Irene told the reporter that science was a suitable career for women, noting that women's abilities were no different from men's. "A woman of science should renounce worldly obligations," she said. Irene also told the reporter that science was the overriding passion of her life but that she also didn't rule out the possibility of having a family one day.

In the mid-1920s, Irene was a rarity, but not a complete oddity. For example, in the United States, the number of women in the field of science was starting to grow, albeit slowly, around this time. Forty-one women earned science PhDs in 1920 versus 138 a decade later. But women averaged less than three doctorates a year in Marie's area of physics. At Marie's lab, Irene would be one of forty-seven women employed there between 1904 and 1934. More than half of those women were foreigners.

Irene was one of the few trained radiochemists in the world, and it's no surprise that institutions began clamoring for a piece of her expertise. Around this time, Irene turned down an offer to teach chemistry for a year at St. Lawrence University in New York State because, as she explained to her friend Missy Meloney, it meant too much time away and wouldn't give her the chance to focus on her research.

Frédéric took advantage of his close proximity to Irene. He peppered her with questions, even waiting however long it took for her to finish work so that he could walk her home. He picked her brain. And in time he discovered that, in actuality, Irene wasn't cocky but confident. She wasn't condescending but commanding. And she wasn't cold-hearted but passionate about people and science rather than money and things, just like her parents. She also was charmingly blasé when it came to fame. Once, when Frédéric asked her what it was like to be the daughter of such famous parents, Irene looked as if she didn't know what he was talking about, remarking that it really didn't affect her.

Rarely have there been two sets of personality traits and interests so completely different. With dark hair and eyes, Frédéric was a charming man, a magnetic character who loved women and people in general. He was an extremely well-rounded athlete, an avid skier, sailor, hunter, fisherman, and tennis player. He also played the piano. He was the type of man who never met a stranger. He loved little more than spending time with friends, enjoying a few laughs. Abundantly informal, those at the institute began calling him Fred even while maintaining the strictest of formalities with everyone else there. In contrast, Irene was gruff and despised small talk. She paid almost no heed to social niceties, believing they only got in the way of her work. She maintained strict standards both for herself and for others, and she never failed to successfully complete any task assigned to her, no matter how daunting. She didn't hunt or fish, and she showed little interest in music.

But the two shared a passion for the underdog. Both were humanists and pacifists. Both were patriotic and loved France. And Irene was an accomplished athlete in multiple sports including skiing.

Frédéric was drawn to her focus, passion, and intelligence, and perhaps by her closeness to the woman he admired most. After joining him on long walks in the countryside, Irene began to succumb to his charms. She grew to appreciate Frédéric's own acumen, as well as the background that led to his sympathy for the working class. Irene's grandfather, although he hadn't fought for the Commune, was a socialist sympathizer. Like Irene, Frédéric had lost his father when he was a child. He had also suffered the loss of an adored older brother who had been reported missing in action in one of the first battles of World War I. In the way they worked, Frédéric and Irene were very much like Pierre and Marie. The Curie women enjoyed focusing more on chemistry while Pierre and Frédéric preferred physics. Like Marie, Irene took her time processing ideas while Frédéric was quicker on his feet. When writing of Frédéric to her friend Angèle Pompéi, Irene said that: "We have many opinions in common on essential questions."

Irene and Frédéric's daughter, Hélène Langevin-Joliot, talked about her parents' disparate personalities. "My mother was rather quiet and very studious and my father was very charming. . . . [B]oth were intelligent. . . . [T]heir collaboration was very successful because their personalities were so complementary," she said.

Frédéric wrote later of his early days with Irene that he hadn't the slightest inkling that they might one day be a couple: "But I watched her . . . with her cold appearance, her forgetting sometimes to say hello, she didn't always create sympathy around her at the laboratory. In observing her, I discovered in this young woman, that others saw as a little brutish, an extraordinary, poetic and sensitive being who, in a number of ways, was a living representation of her father. I had read a lot about Pierre Curie. . . . I found in his daughter this same purity, this good sense, this tranquility."

There was only one problem with the romance: Marie wasn't quite as smitten as her daughter. Indeed, when Irene calmly informed Marie and Eve of her engagement to Frédéric over breakfast one morning in 1926, both her mother and her sister were apparently floored. As Eve put it, the Curie women's inner sanctum had rarely been invaded except for a few very close friends. But now a man had entered the picture, a gregarious creature bound to turn the tightly knit household upside down.

Despite the disruption in everyday life that Frédéric's arrival on the scene must have caused, a review of letters exchanged between the Curie women during this time reveals no change to the atmosphere in which the family lived and worked.

In the midst of the courtship, perhaps before she even knew it was going on, Marie traveled to Prague to meet with researchers there before moving on to the West Bohemian spa town of Jáchymov, where today one of the spa houses is named after her. It was this town that had graciously provided pitchblende ore to Marie and Pierre in the late 1800s. But the trip entailed many official duties that wore Marie out.

On June 14, an exhausted Marie wrote Irene of the journey's compounding pressures, noting that she was "bewildered by the life that I am leading and incapable of telling you anything intelligent." She wondered "what fundamental vice there exists in the human organism that this form of agitation is, to a certain measure, necessary." She said that Meloney might call it the "dignifying of science." Marie seemed annoyed at her tight schedule when she warned Irene that she might not be able to leave for Paris before June 18.

For her part, Irene wrote several letters to her mother, never mentioning Frédéric, but rather relaying plans to visit L'Arcouest to see how construction was coming along on the house the Curies were having built there.

As the year went on, and even as her relationship with Frédéric was heating up, Irene kept to her schedule without cease, accompanying her mother on a number of overseas trips as promised, serving as both her scientific partner and her close friend. And these trips were no brief interludes. During one tour of Algeria, where many people had received French citizenship for their service during World War I, Marie and Irene gave a series of lectures over a period of weeks. Marie wrote Eve of riding camels with Irene across the impressive dunes. "Certainly, upon leaving, one has a certain regret and desire to see them again—and yet it is only sand," she said.

But their biggest trip came in the summer of 1926—one arranged long before Irene announced her engagement: a months-long excursion to Brazil. For Marie it was the chance to fulfill an obligation to France's minister of foreign affairs, who was hoping for a boost in Franco-Brazilian relations. For Irene it was the chance to reveal the warmer side of her personality in a number of passionate letters to her fiancé.

When Irene was in Brazil, a lovesick Frédéric wrote that he found the laboratory "a fraud" without her presence. "I link myself more with human beings than with things; one can work anywhere. . . . [W]hat gives interest to life in the lab are the people who animate it. And since your departure I find it empty," he wrote, according to biographer Rosalynd Pflaum and others.

Irene, too, wrote Frédéric of her love for him and how she longed "to embrace him" upon her return and to run her hands through his hair until it stood on end. Showing her disdain for smoking, one shared by her mother, Irene also asked in letters if he was smoking an "unreasonable" amount of cigarettes. "Try not to smoke like a factory smokestack, speed at a hundred kilometers-an-hour on your motorbike, and do all the other unreasonable things of which you are capable," she wrote him.

Despite Irene's separation from her betrothed, from all accounts the two women seemed to enjoy their time together in South America, and the trip wound up being more pleasure than business. Marie wrote Eve about lecturing at several conferences as Irene successfully performed the accompanying experimental demonstrations. From the start, Marie had made clear to the Brazilians that she would attend only a minimal number of official dinners and absolutely no lunches. Instead, mother and daughter took long hikes in the rainforests, which Marie described as very different from European forests, with their abundance of climbing plants. Marie also wrote that she bathed in the sea for hours at a time in an effort to improve her health.

The lengthy journey, which took two weeks of travel time each way, had been planned long in advance, and Marie saw no reason not to take Irene, even after she was informed of Irene's engagement.

Back in France, Marie remained skeptical that Irene's relationship with such a completely different personality would last. She was concerned about Frédéric's abundant charm and good looks. And she was concerned about his youth. In the end, she insisted that the pair sign a prenuptial agreement stipulating that Irene alone would inherit control of the institute's radium and radioactive substances.

The two women had returned aboard the SS *Lutetia* on September 12 and within a week Irene was back with Frédéric at L'Arcouest. In a letter to her mother, Irene wrote about the prenuptial agreement, saying that she had met with an attorney to discuss a "separation of wealth." It seems Marie had some concerns, too, about property owned in Brittany. Irene wrote that to put the land solely in Marie's name "would run up some considerable costs."

Despite her mother's unease, an undeterred twenty-eight-year-old Irene married twenty-five-year-old Frédéric at the City Hall in Paris on October 9, 1926—the two changed their names to Joliot-Curie and settled into their own flat. The decision to marry Frédéric is evidence of Irene's independent streak. No matter how close Irene was to her mother, she was a strong woman who ultimately did what she wanted. Not unlike Pierre and Marie, the pair went straight from a lunch celebrating their wedding to an afternoon of work at the laboratory. Their honeymoon consisted of nothing more than a week alone in Marie's apartment while Marie and Eve were in Denmark on a lecture tour.

Marie had been so worn out by the wedding that she stipulated to her host, the Danish physicist Martin Knudsen, that she would only accept

dinner invitations from either himself or Niels Bohr, who had won the
Nobel Prize in Physics in 1922, and that she would decline any lunch
invitations no matter who they were from.

Frédéric and Irene enjoyed only a brief period of quiet marital bliss as
Irene became pregnant quickly and Frédéric soon underwent an appen-
dix operation. Although Irene had contracted tuberculosis during World
War I, she gave birth to Marie's first grandchild, Hélène, on September
17, 1927. (Irene worked right up until lunchtime, after which Frédéric
took her to the hospital, where she had the baby in the afternoon.) Irene
seemed to relish motherhood, telling one friend that if she hadn't had a
child, "I would have never forgiven myself for having missed such an astonishing experience." Although doctors confirmed Irene's tuberculosis—
and ordered her to rest and not think of ever having another child—she
went back to work immediately with Frédéric. Marie's attitude toward
her son-in-law softened, and she even began to admire his intellectual
prowess. She practically ordered him to take his second baccalaureate and
then to obtain his doctorate. At one point she conceded to her scientist
friend Jean Perrin that "the boy is a fireball." Marie started to see that she
not only had a daughter off whom she could bounce scientific ideas, but
a new son as well. In no time, the Joliot-Curies were joining Marie for
meals at least four times a week. According to Irene, Marie and Frédéric
often debated with such ardor that she "couldn't get a word in and was
obliged to insist on having a say when I wanted to express an opinion."

Irene's marriage resulted in Eve's spending a lot more time with her
mother. Although at this stage Eve had her own car and an apartment
on the nearby Rue Brancion where she entertained her artistic friends,
she still spent the nights and kept clothes in Marie's home. Over a quiet
dinner together, Marie would go on about all the minutiae of her long
day at work. And Eve was a good listener. She wrote that these evenings
gave her an intimate portrait of what went on at the institute—a world
Eve had never been privy to but to which Marie and Irene belonged,
body and soul. "Apparatus which Eve was never to see became familiar
to her—familiar like those collaborators of whom Marie spoke warmly,
almost tenderly, with the aid of many possessive adjectives," Eve wrote
in the biography of her mother. Marie often would be rambling on about

the laboratory when she would suddenly stop in the middle of a sentence and insist that her daughter tell her about something of interest happening in the world. Eve generally agreed with her mother's political opinions, which ran firmly against dictatorship. If anyone ever had anything good to say about a dictator, Marie was known to always calmly interject: "I have lived under oppression. You have not. You don't understand your own good fortune in living in a free country."

Fortunately, Marie's conversations with her daughter typically turned back to lighter subjects. Eve would often deliver snippets of news of little Hélène to Marie. "Stories about her granddaughter . . . ," Eve wrote, "a quotation from the child's talk, could make her suddenly laugh to the point of tears, with an unexpected laugh of youth."

If Eve were going out after dinner, her mother would often watch her get ready. Marie still had the fashion tastes of a country priest, but she tried to be encouraging in her own way. According to Eve, Marie—who spoke to her daughters in French—would say: "Oh, my poor darling! What dreadful heels! No, you'll never make me believe that women were made to walk on stilts. . . . And what sort of new style is this, to have the back of the dress cut out? You wear black too much. Black isn't suited to your age."

During these evenings, the makeup box always turned out to be a major point of contention. After Eve had put on her makeup, Marie would say to her: "Well, of course, I have no objection in principle to all this daubing and smearing. I know it has always been done. In ancient Egypt the women invented far worse things. . . . I can only tell you one thing: I think it's dreadful. You torture your eyebrows, you daub at your lips without the slightest useful purpose." Marie would tell Eve that she liked her better when she was "not so tricked out." She would promise to kiss her the following morning before she had a chance to "put those horrors on your face." In contrast to Eve, Irene was—even as a married woman— still very much like her mother, seeming to abhor any expenditure of time, money, or energy on her appearance.

Marie engaged in some self-analysis, it seems, during her long talks with Eve, admitting the obvious—that her life had not been like everyone else's. Perhaps she was reflecting on the fact that she'd been away from home so often. Marie told her: "I have given a great deal of time to science because I wanted to, because I loved research." What she wanted for young women, Marie added, was that they combine a simple family life with work that interested them. Langevin-Joliot, also a physicist, said her mother and grandmother probably wouldn't have been very happy if she'd chosen only to become a wife and mother. "Maybe [my mother Irène] would have tried to understand but I don't think she would have wanted [me not to have a career]," she said.

When it came to Eve, Langevin-Joliot said Marie never tried to force her into a scientific career. She described her grandmother's style of child rearing as one that never sought to impose beliefs but simply encouraged the pursuit of one's passion. She said that neither Marie nor her own parents—despite their many achievements—had ever used "brute force" on their children in an effort to push for perfection.

"My brother and I are both physicists but our parents never pushed that on us. Their point was that we have to choose something interesting to do. It didn't matter if it was related to science," she said.

Langevin-Joliot noted that they also were never encouraged to be the best. "I did not learn this as a child. That is more of an American idea. When I was at school, I was first in many things—physics, sport, math. My brother was just an average kid. My mother always told him not to worry that his sister is doing so well. We did not have to be the best," she said. "My mother never heard that either as a child and I never heard that. It's a dangerous idea. The culture of prizes is an American one and it is invading Europe. I don't care for it. My mother chose to do what her mother did not for prizes but just to be happy. That was the most important thing."

Shortly after Hélène's birth in October 1927, Marie attended the Fifth Solvay Conference in Brussels, the most famous of all the Solvay Conferences. This time, the world's most notable physicists met to hash over the newly formulated quantum theory. The leading figures were Albert Einstein and Niels Bohr, both giants of fierce propriety and convictions. At the time, Einstein was disenchanted with Werner Heisenberg's "Uncertainty Principle"—one of the fundamental concepts of quantum physics—that described people's observations as having an impact on the behavior of quanta—or the smallest units of energy. It was during this conference that Einstein famously remarked "God does not play dice" to which Bohr replied "Einstein, stop telling God what to do." And thus the great debate over the interpretation of quantum theory was borne. Of the twenty-nine conference attendees, seventeen of them were or became Nobel Prize winners. Marie, as usual the only woman in the group, was alone among them in having won two Nobel Prizes in two separate scientific disciplines.

During these years, as Marie watched over the affairs of her two very different daughters and the upbringing of her granddaughter, she became obsessed by a new dream. With her Radium Institute in Paris off and running, she wanted now more than anything to create another magnificent radium institute, a hub of scientific research and cancer treatment, in Warsaw, in the land of her birth. But Poland, although free again, suffered from a battered agrarian economy and a poor infrastructure. Involved in multiple border disputes throughout the 1920s, the country—although enthusiastic about Marie's plans—was ill-equipped to fund an expensive capital project. And so, once again, Marie's very good American friend would have to come to the rescue.

As it happened, Missy Meloney, by this time editor of the *New York Herald Tribune*'s magazine, stopped in Paris to visit Marie in 1928 following an interview with dictator Benito Mussolini in Italy. Meloney's husband had died three years earlier in December 1925 at the couple's country home in Pawling, New York, of tuberculosis—the same disease that had killed Marie's mother and that Irene had suffered from since World War I. (When Irene's blood count was abnormal in 1927, Marie wrote her brother that a planned trip to the mountains would be good for her anemia.) Meloney confided fears to a sympathetic Marie that she, too, had contracted the disease. She also truly missed her husband. William Meloney had been a warm and colorful character, a native of San Francisco who ran away to sea at the age of eleven. Marie, herself a widow, comforted her friend. But then later, as the conversation moved on, a desperate Marie also took advantage of the visit to ask another favor. She needed America's backing—again. And only Meloney would be able to help her get it.

CHAPTER 8

Turning to America—Again

Although Marie was a proud citizen of the Third French Republic, she had never denied her roots, nor lost her sense of Polish identity. Marie was never a Polish citizen, although her parents had been Polish nationalists, since the Polish state didn't exist when she left Warsaw to attend college in France. Yet she traveled back to Poland frequently and still felt an allegiance to the small but plucky nation's politics.

During 123 years of occupation, the Polish people had battled hard to hold on to their history, culture, and language. After the country won its independence in 1918, after World War I, the Poles were eager to reassert themselves on the European stage. By the late 1920s, Marie began working closely with Polish officials on her goal of building a world-class scientific research facility in Warsaw. She had no problem gaining the backing of the government, the city of Warsaw, and many Polish institutions as the country had always been as proud of Marie as it had of other famous Poles such as Copernicus, the father of modern astronomy.

But to build something of significance on the ruins left over after Russian domination—a Radium Institute on a par with the one in Paris—probably meant more to Marie than to anyone else.

Even so, all the enthusiasm in the world didn't mean much when there was so little money to back it up. An unwieldy economic structure put in place by the Russians, too, attested to the steep challenges facing Marie and her sister in trying to get anything even remotely resembling a successful institute off the ground quickly. Marie had called upon the population to give donations for the institute and also had contributed some of the money from her 1921 trip to America for the "renting" of radium for Warsaw researchers. Marie's sister Bronya, doubling as architect and treasurer, had been put in charge of a grassroots effort that in the end enjoyed at least modest success. Posters had been plastered in every corner of the country, bearing Marie's likeness alongside the slogan: "Buy a brick for the Marie Skłodowska-Curie Institute." Buoyed by the support of so many Polish officials, the campaign was able to amass a respectable bunch of bricks after only a few months. If these bricks weren't enough to complete the institute, they were at least enough to throw up a few walls and lay a foundation.

And so finally, on a bright sunny morning in 1925, Marie enjoyed one of the best days of her life when the president of the Polish Republic ceremoniously laid the first stone of the institute, followed by Marie and, ultimately, the mayor of Warsaw. She'd always thought the location of the institute perfect, a lovely symbol of Polish pride standing alongside one of the city's broadest streets and not far from the brand new airport. It was the perfect setting to enjoy a perfect day.

Polish President Stanisław Wojciechowski had known Marie since their student days at the Sorbonne, and the groundbreaking was an informal ceremony. Wojciechowski expressed astonishment at Marie's ability to speak flawless Polish even after so many years in France.

The president asked Marie: "Do you remember the little traveling cushion you lent me thirty-three years ago when I went back to Poland on a secret political mission? It was very useful." She replied, laughing: "I even remember that you forgot to return it to me."

For all the Skłodowska siblings, ever patriotic, the event called them back to the days of their youth, during which their father had instilled in them a patriotic duty that would linger forever.

But it wasn't enough. Even as the bricks slowly turned into walls, one year after the next, it became clear to Marie that, no matter how great her sister's organizational prowess, she would have to solicit substantial outside support if she was ever to supply the institute with the radium needed to provide widespread cancer treatment. Both Marie and Bronya had already contributed a large chunk of their own savings to the institute. But the well was running dry. And Poland, which was seeking to rebuild itself following independence, was tapped out as well.

Perhaps, though, Marie wasn't worried. After all, she could always turn to a virile America for help.

Unbeknownst to Marie, however, America had changed a great deal since her seven-week visit. There was no guarantee this time around that a more inward-minded country would again come to the rescue of a struggling European physicist.

Even so, Marie turned her attention to the West for the second time. When Meloney stopped in Paris during her visit to Europe in 1928, Marie didn't hesitate to ask whether she might be able to work her magic again in order to benefit an institute that meant as much to her, if not more, than the Radium Institute in Paris. Marie told her, with typical bluntness, that she absolutely had to have a second gram of radium and that she wouldn't be able to get her hands on it without the generous support of the people of America. Meloney, who never seemed immune to her dear friend's appeals, immediately began organizing a second Madame Curie tour of America, realizing from experience that the promise of her presence was the only way to secure a large influx of donations. She told Marie: "I no longer find many things in life worthwhile, but to serve in even this menial way in a great cause is a real compensation for me."

But, again, the American people were not in the same frame of mind as they were in 1921. Rather than play superhero, they seemed more inclined to let Europe stew in its own juices. The government, too, had grown so focused on domestic matters that it failed even to join the League of Nations despite the fact that it had been America's idea to create the organization in the first place. By the 1920s the United States had not only raised tariffs on imported goods but also put a stop to free immigration following World War I. And surplus money was in short supply. Elected in 1928, Herbert Hoover was governing on the cusp of the Great Depression, which would carve its trough into the nation's economy in 1929.

Faced with a more distracted America, an uncharacteristically discouraged Meloney wrote to Marie: "The Polish radium seems far off. I have not been well and America is in the throes of a terrible political upheaval." That upheaval was the Teapot Dome, a case of bribery involving Secretary of the Interior Albert Fall. Fall, who had rented government lands to oil executives in exchange for personal loans and gifts, was found guilty of bribery in 1929 The event marked one of several scandals that ruined the reputation of President Harding, the man who had presented the gift of radium to Marie in 1921. Harding had died in office only two years later and was succeeded by Vice President Calvin Coolidge. As the US secretary of commerce under both Harding and Coolidge, Hoover easily won the Republican nomination and then the next presidential election in 1928 despite not having any experience in elected office.

Even the ever-optimistic Meloney wasn't completely certain that—in this environment—Americans would be in the mood to open their pocketbooks to a foreigner again.

Since Marie's last visit to America, Meloney had been one busy lady. In 1926, *The Delineator* magazine had merged with another Butterick Publishing Company publication, *The Designer,* with Meloney continuing at the helm. In her new role, she won kudos for instigating a major conference on food habits sponsored by the US Department of Agriculture and held in April 1926 in Washington, D.C. The large-scale affair came about after Meloney's *Delineator* uncovered and then brought to light the complete lack of any official standards for assessing the weight of adults in the United States save for US Army charts formulated after the Civil War. Meloney enlisted the help of Dr. Samuel Brown, president of the New York Academy of Medicine, who agreed to co-organize the event with an eye toward developing a weight table similar to that in use for children. The thinking was that such a table, based on weight, height, and age, could help thwart the era's unhealthy trend of being fashionably and unnaturally slim. Afterwards, Meloney was named editor of the Sunday magazine of the *New York Herald Tribune,* a paper that boasted

(Right) *Marie Curie alongside her daughter Irene in 1921.*

(Below) *Marie Curie and Marie "Missy" Meloney in the United States in 1921.*

Marie Curie arrives in the United States in 1921 and holds her first press conference on the deck of the Olympic *as her daughters Irene and Eve look on.*

Marie Curie in the United States in 1921.

(Right) *Marie Curie with US president Warren Harding following the White House ceremony.*

(Below) *Marie Curie visits a chemical company in Pittsburgh in 1921.*

Marie Curie receives an honorary degree from Columbia University in 1921.

Marie Curie at a meeting in Geneva of the International Committee on Intellectual Cooperation in 1925 along with Albert Einstein, Robert Millikan, and Gilbert Murray.

Marie Curie at St. Lawrence University in
Canton, New York in 1929.

Marie Curie and Irene
with other researchers at the
Radium Institute in Paris
in 1928.

Marie Curie and US president Herbert Hoover in 1929.

Marie Curie and Irene Joliot-Curie at the Radium Institute in 1930.

Marie Curie attends the Solvay conference in 1927.

Irene and Frederic Joliot–Curie with their children Helene and Pierre, Marie Curie, and Frederic's mother Emilie Roederer-Joliot in 1932.

Irene and Helene with Marie Curie at their family home in Brunoy in 1930.

writers such as Dorothy Thompson and was the forerunner to *New York* magazine.

All of this meant she was rising quickly in the ranks of influential society. Yet Meloney had to be frank with Marie. She begged her not to bring Bronya to America, as was her initial wish, because she was afraid that the American people would not be as magnanimous in their support if the emphasis was on an obscure Polish institution as they had been when the fundraising campaign was framed around the world's greatest scientist seeking radium to research cancer treatments. (In the end, Bronya decided not to come because she didn't want to detract attention away from the mission.)

The good news was that Meloney had no trouble convincing Marie to come to America even though, as before, Meloney had to dole out every assurance that she would be shielded from an insatiable press and from having to meet an unreasonably large number of people. In fact, this time a more physically frail Meloney went along with Marie's demands without much of a fight. Marie would not give autographs or interviews, sit for portraits, or attend large dinners or receptions except for a few unavoidable ones, including an event honoring Thomas Edison. She wouldn't even shake hands—with very limited exceptions. Indeed, in the end, she was willing only to make four public appearances during her entire stay. No matter. Meloney was relieved just to be able to guarantee her presence.

Despite her initial reservations, a pleasantly surprised Meloney was able to quickly raise the $50,000 needed to purchase a gram of radium— which had dropped in price since 1921—once Marie's visit was publicized. This meant that Marie would no longer have to spend the $3,500 in yearly annuity she was receiving from the $50,000 left over from her 1921 trip to rent radium for the Warsaw institute. She would now be free to use that yearly income as Meloney and the other fundraisers had originally intended—for her own personal needs.

Meloney also encountered no pushback when arranging for Marie to be an official guest of the White House, effortlessly securing an invitation from Hoover, a man Meloney held in high esteem. Perhaps this was because Meloney was such a staunch Republican. She had written many letters to Marie about him, portraying him not as a typical politician but as a humanitarian who cared about the sciences. Hoover, an engineer, had sat on the Marie Curie Radium Committee in 1921 and had met Marie during her first trip to America—although Marie reportedly didn't remember this. His invitation for her and Meloney to stay at the White House was a first as no foreigner had ever been given such a privilege.

And so plans were in motion for what would hopefully be a second successful tour.

Marie's daughters were so busy with their own professional obligations that they were not able to accompany their mother on this second trip to America. In addition to her own research, Irene was overseeing the work of a new crop of Sorbonne students employed by Marie at the institute and she also was managing to run the household even though she never learned to cook. As a result, Meloney secured another companion for Marie. Lucy Goldschmidt Moses, or Mrs. Henry Moses, was a well-known philanthropist from New York and a member of the Marie Curie Radium Committee in 1921. She was only too happy to accompany the great scientist on her journey.

And so in early October 1929 the two boarded the SS *Île de France* and headed for America. The captain's concern about his famous passenger's well-being was so great that he proclaimed his intention to change course if he encountered even a hint of rough waters. Fortunately, Mother Nature cooperated and it was smooth sailing; Marie arrived in New York without ever having gotten sick. Even so, she remained perpetually fatigued, which led her to send word ahead that she did not wish to be photographed or interviewed upon her arrival.

This time around, Marie was sixty-one, bent, pale, and white-haired. And so, after the ship docked on October 15, 1929, instead of confronting a mess of reporters, she slipped down a back stairwell to the lower level of the pier where a shiny limousine waited to whisk her to Meloney's apartment.

As the *New York Times* reported, "Representatives of three Governments welcomed her, cut the red tape of landing, shielded her from annoyance and treated her as if she were a little old lady made of glass." The article titled "Mme Curie Arrives, Happy to Be Back" made no mention of Poland until the very last bit of the article, reporting, "This gram of radium Mme Curie will donate to the Radium Institute under construction at Warsaw." In other press coverage of her trip, the fact that the radium would go to Poland was either not mentioned at all or was buried somewhere at the end of articles, probably at Meloney's encouragement.

As during her first trip to America, Marie spent a great deal of time between her public duties either relaxing at Meloney's Fifth Avenue apartment in Manhattan or at her sprawling country estate in Pawling, New York.

Marie's letters to her daughters during the American visit never stopped, one week after another, although they arrived sporadically and were often out of date.

During her first trip in 1921, Marie had been so exhausted by her myriad of social obligations that several events had to be canceled. Although the 1929 visit was designed not to be as cumbersome, it still seems to have worn Marie out—and right from the start.

In a touching letter to Eve written on October 20, just five days after her arrival, Marie said how much she hated not only being away from her daughter but also being made to feel like an object of curiosity.

Some things, it seems, never changed.

"I am seeking, however, to surmount my extreme repulsion for this whole situation and keep my eye on the real issue which is to gather the necessary funds to allow me to pursue my work," she wrote. "But it is obviously not possible to do my scientific work here living the way in which I have been."

She told her daughter to think of her sympathetically "and be sure that I am trying to take things positively and not allow myself to become overwhelmed with worries."

Never one to downplay the kindness of her friend, Marie also noted that she found "it very touching that Mrs. Meloney has so much affection for me and also that she desires to help me with my work."

Yet she couldn't resist making a complaint: she despised having to stay with strangers and exchange banalities. "There's packing up and unpacking and looking for one's things and also vainly trying to seek a little repose and peace," she wrote. "How can I cope with this regime with serenity I wonder.

"I am also extremely annoyed that the trip to the White House has been postponed until after my return from Detroit and the University of St. Lawrence because a week dedicated to this expedition will be very strenuous and I am afraid that I will be exhausted for the ceremony in Washington, which is what I must be in top form for," she said.

Spending a night in the Long Island mansion owned by Nicholas F. Brady, a major player in the finance sector, and his wife Genevieve, Marie found herself face-to-face with original Whistler paintings, Waterford candelabras, and three hundred meticulously landscaped acres, all of which made for a memorable stay.

Even more memorable was the journey to get there.

After leaving Meloney's home by the service entrance in order to avoid sixty reporters huddled outside the front door, Marie had a wild ride to the Brady compound, which she described to Irene.

She wrote: "I came down the service stairs to avoid sixty reporters waiting at the main entrance. Then we made a sensational ride from New York to Long Island, with a policeman on a motorcycle in front of us sounding blasts on his siren and scattering all the vehicles on our road with an energetic movement of one hand or the other; in this way we carried on like a fire-brigade off to a fire. It was all very amusing."

Marie also told her daughter that she was about to leave for Detroit aboard a train accompanied by a "first-rate doctor."

"He gave me an inhalation of pulverized oil to have now and again," she said. "If that could replace, from the point of view of the cough, the odious treatment that was recently given to me in Paris where they put oil in my trachea with tweezers, I would be delighted."

One of Marie's first and most satisfying stops on this second American tour was in Dearborn, Michigan, to attend an event on October 21, celebrating the fiftieth anniversary of Thomas Edison's invention of the electric "lamp." If nothing else, it gave Marie a firsthand look at the changing scope of American-style public relations. Although the country was exceptionally skilled at promotion, by the late 1920s at least some had started to view sensationalistic publicity as a menace to the integrity of the press. With that in mind, Edison's company, General Electric, hired Edward Bernays, a pioneer in the field of public relations, to oversee the press surrounding the fiftieth anniversary of the first incandescent light. Bernays welcomed the opportunity as a chance to prove that public relations was indeed an honorable profession. He was the perfect man for the job. Over the years he would sell America on everything from presidents to Ivory soap.

Six months before Marie's arrival, Bernays launched a massive publicity campaign called "Light's Golden Jubilee" that honored both Edison and his bringing down of "light from the heavens." If nothing else, Bernays knew how to grab people's attention. In dramatic fashion, gaudy light displays arranged in cities across America and in a number of foreign capitals were set to be dimmed—all at the same time—before being slowly brought back to full brightness in honor of Edison. Ultimately, the Postmaster General even issued a commemorative stamp.

But the high point was the lavish affair in Dearborn, Michigan, attended by Marie along with more than five hundred other scientific and political dignitaries. For once, Marie was delighted to take part in an extravagant gala, enjoying a stay in the home of her friend Henry Ford, who had turned his fledgling idea of a motorcar into a reality. Edison was Ford's role model and, as a young man, one of Ford's first jobs was at the Edison Illuminating Company, where he worked his way up to chief engineer. Indeed, he so admired Edison that he actually reconstructed Edison's invention factory in Menlo Park, New Jersey, as a museum in Dearborn.

In addition to Ford and Marie, other guests included Orville Wright, J. P. Morgan, Will Rogers, and John D. Rockefeller Jr. Every wire service and important media outlet covered the tributes paid to Edison that evening, tributes that were broadcast to the world by radio.

Marie found such dinners in America to be a lot different from dinners back in Europe. She wrote her daughters that the event attracted five hundred people—"a lot for me but not a lot here."

"The dinner was composed of only a few courses, no wine or liqueurs, and very few speeches," she wrote. "It seems to me that the big industrialists are not as loquacious as the university lecturers."

She enjoyed the evening even if the man of the hour—Edison—was in seriously poor health.

Marie described him as extremely elderly and able to recite only a few words. She also oddly mentioned that she had dined before going to the Edison dinner "because one doesn't expect to eat at that type of gathering." She described the dining room as being candlelit until, suddenly, the electric lights all came on at once in a flamboyant fashion. Edison's health had deteriorated to the point that he couldn't stay up for the entire evening. Less than two years later, Edison lapsed into a coma before dying from complications related to diabetes at his estate in West Orange, New Jersey.

But being surrounded by so many important figures, all coming together to celebrate Edison, brought Marie enormous satisfaction, and she was glad she had made the trip. She also praised Meloney for the way she refused, on Marie's behalf, any and all requests for autographs. The attention to detail had made the evening a pleasure.

After the Edison affair, the next stop on Marie's tour was Schenectady, New York, where, on October 23, she toured General Electric. So honored was the company that it halted all operations at the plant during the visit so that everything—every piece of equipment—could be placed at her disposal. Only once before had General Electric ever given someone carte blanche in this way—and that was when Charles Lindbergh visited the plant.

While she was there, Marie was invited to conduct any experiment she desired, no matter what it was, and to examine or use any piece of apparatus she fancied, with the help of all the scientists she wanted. Marie was dumbstruck by the enormity of the enterprise, with the endless rows of identical high-precision instruments moving along massive assembly lines. Everywhere she looked there were new kinds of machinery being handled by an army of dedicated research workers—none of whom were female.

During her time there, she was hosted by yet another brilliant businessman, Owen Young. Young had become president of General Electric in 1922 before eventually being named chairman of the board. Young was not only one of Marie's financial advisers, but he had also donated a lot of equipment to her institute. Born in 1874, Young was sixteen years old when his parents mortgaged the farm to send him to St. Lawrence University. Upon his graduation in 1894, he turned to Boston University where he completed the three-year law course in two years while supporting himself by tutoring. He became a lawyer and caught the attention of Charles A. Coffin, the first president of General Electric. Later, under his direction, General Electric began the extensive manufacture of electric appliances for home use.

It was only fitting then that, after the General Electric tour, Marie agreed to dedicate the Hepburn Chemistry Laboratory at St. Lawrence University in Canton, New York, which had a carved figure of her perched at the building's stone entrance. She had been invited to dedicate the building in 1926 but, since she couldn't attend at that time, the university held off on the dedication until she could make the trip.

Marie spoke out in favor of pure science as the source of true progress and civilization, especially after witnessing how much America's burgeoning scientific environment was intermingled with trade, politics, and even war debts.

"I also believe that pure scientific research is the true source of progress and civilization and that by the creation of new centers the number of men and women who are able to devote themselves to science shall be increased," she said.

No doubt there was part of Marie that was put off by America's flash, but there was another part that admired its youth and vitality and, also, the way profits from mass production were being reinvested in research laboratories.

Revealing the university's admiration for its guest, the oldest member of the faculty, Dr. Charles Kelsey Gaines, from the class of 1876, composed a sonnet that honored her accomplishments.

What age-long effort has essayed in vain
  This woman wrought. She loosed the Gordian knot
  That held the conquest of the world, and what
  The frustrate alchemist could ne'er attain
  She has achieved. She broke the primal chain

That binds the elements; she touched the spot
Where lies the hidden spring,—and lo! the plot
And secret of the universe lay plain.
And what the alchemist in vain had sought
For greed and dazzled by the lure of gold,
She only that she might the truth unfold,
Still toiling for the love of man has wrought.
Let all the ghosts of alchemy bow down
While on this woman's brow we set the crown.

After the ceremony, Marie was asked to plant a lovely symmetrical evergreen outside the building to become "a living memento of her visit to St. Lawrence," according to writer Denise Ham. Amusingly, she was handed a small souvenir shovel, which she was supposed to manipulate to pretend she was digging in a show designed for the press. Instead, Marie tossed it aside, grabbed a real shovel, and began digging a real hole, all by herself. Everyone was stunned by the older woman's enthusiasm for the menial task. "I do this very willingly, and hope that your university will grow as the tree," she said.

Following this visit, Marie rested briefly at Meloney's country home before heading with her to Washington on October 29 for the main event. Compared to the extreme stress of the earlier visit, this trip to the White House would be relatively carefree. Arriving at Union Station, the two friends were met by the president's military aide, who escorted them to the White House, where they enjoyed a relaxed dinner with the president and his family.

In a letter, she told Eve about a little ivory elephant she was given as a gift.

"It seems that this animal is the symbol of the Republican party, and the White House is full of elephants of all dimensions," she said.

Her big prize came the next day. The radium worth $50,000 was presented to her on October 30, 1929, in a building of the National Academy of Sciences and National Research Council, a much less glamorous space than the East Room, the site of the radium presentation eight years earlier. The gift itself was in the form of a bank draft encased in silver that was, again, less formal than the golden key handed to her in 1921.

This time, in order to keep costs down, nations had been encouraged to place bids. As a result, the actual purchase of radium was made in Belgium, the country with the lowest bid.

When it came to asking Americans for money, Marie couldn't have been luckier in her timing. Only the day before the White House presentation, on October 29, the stock market experienced its worst decline in history when 16.4 million shares of stock were traded during what would be dubbed Black Tuesday.

Despite all this, though, Hoover had decided to go ahead with the presentation and enthusiastically paid tribute to Marie. He even paid tribute to Poland.

He said: "In the treatment of disease, especially of cancer, [radium] has brought relief of human suffering to hundreds of thousands of men and women. As an indication of the appreciation and the respect which our people feel for Madame Curie, generous-minded men and women under the leadership of Mrs. William B. Meloney have provided the funds with which a gram of radium is to be purchased and presented to the hospital and research institute which bears her name in Warsaw. The construction of this hospital was a magnificent tribute by the city of her birth and the Polish people, in which the American people are glad to have even this opportunity of modest participation."

She responded that the American gifts of radium were symbols of the country's friendship with France and Poland.

But she also expressed her conviction that "scientific research has its great beauty and its reward in itself and so I have found happiness in my work." The bonus, she noted, was "an additional happiness to know that my work can be used for relief in human suffering."

The next day, October 31, Marie was back in New York at a banquet for the New York City Committee of the American Society for the Control of Cancer, a group headed up by Mrs. Robert Mead, one of the driving forces behind the Marie Curie Radium Fund. By this time, Marie was so tired that she appeared to be in a trance. Thankfully, her formal obligations had been fulfilled. Although Marie had complained of a nagging cold to Eve in a letter dated October 21, she also talked about the great friends she had in America, noting that "in one way I regret that this distance is too large to promote easy communication . . . and too large to come here easily from Europe as well as to get around the interior of the states itself."

Marie's second tour of America was hampered by much stricter conditions—imposed by Marie herself—with the scientist much more focused on the business at hand than on making a good impression, at Meloney's urging. At one point, a somewhat bemused Marie wrote Irene about the differences in custom she had encountered this time. When she found that the bedroom doors had neither keys nor keyholes, she complained to Irene that Americans "come into your room sometimes having knocked, other times not taking the trouble. The maid thinks that she pleases me by moving my things around, displacing them, opening and closing the windows ten times in an hour without consulting me, turning the fire when there's no need. I spend my time keeping her from making me happy according to her personal views. My good friend Mms Meloney has the same tendency, but it's rather touching the way she tries hard not to annoy me."

She stayed at Meloney's home for another week or so, seeking to counteract the well-intentioned desires of Meloney and the maid to make her comfortable at every turn. On her birthday, November 7, she received hundreds of presents from Americans she didn't know: flowers, books, objects, checks for the laboratory and even a galvanometer, some ampoules of radon, and some specimens of rare earth. Turning sixty-two, Marie celebrated by motoring with Meloney through a brisk Central Park before stopping off at the J. P. Morgan Library, a museum and research library founded in 1906 to house the private library of J. P. Morgan.

A mollified Marie sailed back to France a short time later, leaving behind an America just coming to grips with its financial troubles. Marie would never again return to the United States, which was about to enter its Great Depression, with the stock market crash being just one contributing factor. Hoover, the man Meloney placed on such a high pedestal, believed the government shouldn't intervene in the economy, arguing that families could turn the economy around if they just continued to work hard. In June 1930, he signed the Smoot-Hawley Tariff, which increased the tariff rates on imported goods, causing foreign nations to boycott American products in response. All of this, coupled with a lack of banking regulation, led to a massive economic downturn that placed millions of Americans out of work.

Somewhat amusingly, Marie had written Eve just before leaving New York that she believed the country's financial troubles were calming down.

"The government and the banks which bought large numbers of shares have probably made a good deal as well as stopping the panic," she wrote. "I found it very tiresome to listen to all of that talk and I would have preferred not to have been here during this crash if I could have avoided it."

But Marie had accomplished the purpose of her trip to America and was armed with enough money to buy a gram of radium for her institute. She had also come away with many gifts of radioactive elements and scientific equipment as well as promises of scholarship endowments for her institute.

When it came to her Warsaw institute, Marie was about to ride a new wave of prosperity with an infusion of funds and energy.

CHAPTER 9

Into the Spotlight

Eve wrote Meloney in late 1929 that she'd never seen her mother in such good spirits as when she returned from America—although her happiness would be short-lived.

Marie was basking in the glory of her Warsaw institute's potential as the infusion of US funds began plugging holes in her budget. In addition, just after her return, in late 1929, a talented Curie scholarship recipient named Salomon Rosenblum made the first major discovery at Marie's laboratory in Paris when his work with a rare radioactive element known as actinium (prepared by Marie herself) revealed that the alpha rays emitted from a nucleus don't all boast exactly the same kind of energy—a conclusion that helped confirm the new atomic theory known as quantum mechanics. By this time, Marie's laboratory was becoming a bustling center of research, with nearly forty researchers besides Rosenblum (compared to only a handful right after the war).

But it was Irene and Frédéric, working as a team, who were about to become the real stars of the show with an even more dazzling discovery. Throughout history their work has often been overshadowed by the research performed by Marie and Pierre. Yet the younger couple's discoveries were about to help solve the problem of how to release energy from the atom, changing the world forever. Indeed, their soon-to-be-famous discovery would—in the near future—further confirm the instability of the atom, marking a first step toward the harnessing of atomic energy and the development of the atomic bomb in the United States in 1944. Even early on, both Irene and Frédéric were well aware of the destructive potential of nuclear fission.

In many ways, the pendulum of Marie's own life swung slowly during the years following her second trip to America as she increasingly took on the duties of "teacher" and "manager." She chipped away at her definitive book on radioactivity, eventually published in 1935, and gave regular lectures on radioactivity every Monday and Wednesday, in addition to visiting her beloved Radium Institute in Paris on work days.

Thanks to American support, her second facility—a 120-bed Radium Institute—would open in Warsaw in 1932 with a gram of radium at its disposal. Mrs. Andrew Carnegie even sent Marie additional funds, invested in French bonds, with the aim of restoring to its original level the depreciated income of the scholarships established by her late husband. Other American women made superfluous contributions as well, even purchasing dogwood trees for Bronya to plant in the spaces between the four buildings of the Warsaw institute. And Marie received an extra bonus. As Meloney and her fundraising committee had originally intended, Marie was finally able to take for her own personal use a $3,500 yearly annuity from the surplus account set up with the balance of funds raised during her 1921 trip. Previously, Marie had been forced to tap into those funds to rent radium for her researchers in Warsaw.

With her finances in order, Marie spent more time mentoring during her final years, with young researchers repeatedly intruding on her time.

"Is Madame Curie here yet?" "I'd like to see Madame Curie. Has she come?" "Have you seen her?" "Is she here?" "Is she here?"

Every morning, five, ten, a dozen, or even more researchers would assemble near the entrance of the laboratory at the Radium Institute in Paris, a collective wanting of answers and encouragement.

Marie affectionately called them "the Soviet," these young men and women dressed in white coats, all of whom she treasured and thought of as "her children" even if their questions, at times, made her weary before the work day began. For these aspiring scientists a few words with *la patronne* would have been enough to keep them going the rest of the day.

The good news for the Soviet is that it never had long to stand around. One could almost set one's watch by the opening of the gate on the Rue Pierre Curie each morning. At 9 A.M., the car that was a gift from Henry Ford would pass through, the iron door clanging, and the petite frame of Madame Curie would emerge at the garden entrance. The clutch of soliciting young people would rush enthusiastically toward her, making sure they never came too close for her comfort.

Bent on getting her attention, one would say "I've prepared a certain polonium solution . . . if you could just have a quick look," while another, with a hint of uncertainty in her voice, would interject "I've just completed a measurement I'm not sure about. . . ." Fortitude and patience were called for—both qualities Marie could claim. And although she sometimes complained about the commotion, she also luxuriated in the possibilities presented by all the experiments going on around her. And so rather than immediately retreat to her own office, she would often linger, without even removing her coat, liberally doling out advice in a scientific vernacular that would have sounded like a foreign language to any lay person.

These encounters carried on for some time as more workers amassed and the Soviet ballooned into a hive of animated conversations. They were a disparate group, tied together by their scientific proclivities. About a dozen of the researchers were women, an incredibly high number for the time. Many were foreigners from Spain, Poland, Romania, Portugal, Yugoslavia, and even Palestine. (However, it took Marie at least fifteen years following World War I to even consider hiring a German—male or female—at her Radium Institute.)

Marie had returned from America not only with a new enthusiasm but also with gifts for her entire family that she had purchased herself— an unusual gesture by the thrifty matriarch—including new boots for Frédéric.

Like Irene, Frédéric was making strides in his education, successfully defending his own doctoral thesis, "A Study of the Electrochemistry of

Radioactive Elements," on March 17, 1930. Again, Marie hosted a celebratory tea in the garden outside the institute. The thesis was judged to be so superb that he was awarded one of the few new government grants for scientists, one so substantial it allowed him to devote all his time to research. If not for this grant, in order to support his family Frédéric probably would have sought a job as an engineer, a lucrative profession but one that would have been much less to his liking.

By this time, Eve, too, was making her own mark in the world—the world of the arts—as she emerged from her famous mother's shadow and gained prestige on her own terms. However, after giving piano recitals in Paris, in other cities throughout France, and in Belgium, she conceded that she had started music too late in life to make a good living at it. And so, after seeking Meloney's counsel, she put her knowledge of the arts to other uses and began to write music reviews in the weekly journal *Candide* under a pseudonym to avoid trading on the family name. Next she branched out as a film and theater critic for various Paris magazines. She also began translating into French a Broadway melodrama named *Spread Eagle* by George Brooks and Walter Lister about an American businessman who orchestrates a US invasion of Mexico. In addition, she looked after her mother's house, and even her mother's wardrobe, making certain Marie wasn't completely unstylish. But at times she must have felt the sting of being the outsider. At the occasion marking the twenty-fifth anniversary of the discovery of radium, one of the scientists who spoke, H. S. Lorentz, made mention of the promising collaboration taking place between Irene and Marie, adding that "I am quite sure that Mlle Eve Curie would like to do this as well as her sister, but in the end we can't all be physicists."

Although this period was in many ways a prosperous one for the family, the Curies and the Joliots were never to escape heartache. And France was about to enter a tumultuous period.

Soon a string of misfortunes befell Marie's inner circle. When, in 1930, she heard that Pierre's brother was extremely ill and had been in bed for three weeks, she raced to his home in Montpellier to be by his side, despite her own declining health. And she was overcome with sorrow when she heard the news that one of Bronya's children, a daughter named Hela, had died in a possible suicide in Chicago.

According to biographer Susan Quinn, she wrote an American friend that: "We have had a deep family grief because my sister has lost, by accident, her only daughter. . . . I loved that girl." Bronya also lost her beloved husband, Casimir, around this time. Marie had been extremely close to Casimir, who had helped Bronya look after a malnourished Marie during her early days at the Sorbonne.

In October 1930, Marie wrote Eve that she was worried about Bronya: "What I fear most for her is the moment when she must step down from her activities because then she could let herself get overwhelmed by the past that haunts her."

Now an elderly widow with two grown children, Marie began to feel a bit sorry for herself with none of her siblings close by.

"There are three of you in Warsaw," she wrote Bronya in 1932, "and thus you can have some company and some protection. Believe me, family solidarity is after all the only good thing. I have been deprived of it, so I know."

In the early 1930s, scandals related to radium also continued to weigh heavily on Marie's mind. One of the most infamous cases involved a millionaire convert to the benefits of radium named Erben Byers, a national amateur golf champion in America. After he hurt his arm, a doctor prescribed as an energy booster a patented drink named Radithor, distilled water containing traces of radium and mesothorium (another radioactive substance). The drink was touted as a treatment for fatigue, joint pain, and muscle aches. But after drinking several bottles a day for five years, he died at fifty-one on March 11, 1932, in a New York City hospital, a ninety-two-pound shell of his former bulked-up self. An autopsy cited radium poisoning as the cause of death. When asked about such incidents, Marie always reminded people that she wasn't a doctor, although by this time she never dared deny that there were dangers associated with prolonged exposure to radium.

Clearly Marie was worried about the growing use of radioactivity in consumer products. In September 1932, Eve wrote to Marie that she had consulted an attorney regarding the advertisement of a cream called Tho-Radia launched by a French doctor named Alfred Curie (no relation). Apparently Eve and her mother wanted no part of the cream and wondered whether an Alfred Curie even existed.

But Alfred Curie was real and had created an entire line of beauty products that included creams, makeup, and even toothpaste. Tho-Radia contained thorium chloride and radium bromide, both of which were radioactive. The line was evidence of the craze for radium products that went on in the 1920s and 1930s, a time when radium was added to everything from chocolates to cigarettes.

But perhaps Marie's biggest worry around this time was Irene's health, which was deteriorating month by month. Irene had never seemed to fully recover her strength after the birth of Hélène in 1927. After being diagnosed with tuberculosis, a bacterial disease that typically attacks the lungs, Irene suffered from a chronic cough, fever, and weight loss. As was the family's custom, no one talked publicly about how Irene was feeling. Like Marie, Irene's habitual reaction to her poor health was that her symptoms would go away on their own if she simply ignored them. But they didn't go away, and her fatigue grew so severe that, as Marie wrote, Irene finally agreed to seek a cure by going away to Savoie in the French Alps in July 1930. Marie wrote Eve of her delight that her granddaughter, "her little darling," was finally getting to enjoy some quality time with Frédéric while Irene was away recuperating. The truth was that, more and more, Frédéric would be forced to juggle the household, children, and laboratory work as Irene was often off seeking to regain her health in one far-flung place or another. Marie wrote Eve optimistically that "there hopefully will be a good result in Irene's treatment that will soon be announced." But Marie seemed more pessimistic in her writings to Meloney. Thinking Irene was seriously ill, Marie wrote that Irene ought to immediately stop work for six months or longer. But Irene was as stubborn as her mother and wouldn't hear of cutting back. And so, by late in the summer of 1930, Irene returned home and was deemed well enough to go to Brittany with Frédéric and their three-year-old daughter. In her suitcase were the findings of various laboratory experiments conducted in her absence, all of which she would review.

Irene wrote often to Marie from Brittany, proclaiming high hopes for Marie's health going forward.

"Make sure that it is the same for you and, to that end, endeavor to give yourself a break from your worries," Irene wrote. "I am thinking about the periods when you do not have to make any decisions."

"You should abstain as much as possible from thinking about troublesome things in order to rest your mind," she added.

Although a doctor told Irene that her tuberculosis could negatively impact a new pregnancy and worsen her own condition, Irene went ahead and got pregnant anyway, giving birth in March of 1932 to a son. They named him Pierre, a move that thrilled both Marie and Pierre's brother. Like Marie and Pierre, Irene and Frédéric continued to spend long hours at the laboratory even after the birth of their second child, but they also cherished their family time—as well as any time spent outdoors. No matter how busy they were with work, they always went ahead with their vacations, especially those spent at their home in L'Arcouest. They seemed more than willing to swap their research for hiking, swimming, and tennis when the time came. As in her professional life, Irene held high standards when she traveled. In a letter to her mother written during a vacation in France in 1931, Irene complained of the hotel. "As to the level of comfort in hotel rooms in this country, it is better not to talk about it," she wrote.

For a woman who was often away from home when her own two daughters were little, Marie was a doting grandmother and, in a letter written to Irene and Frédéric in December 1929, she wrote about how moving it was for her to observe the evolution of "this little creature who expects everything from you with unlimited confidence and who certainly believes that you can interpose between her and all suffering."

Marie also wrote how, one day, the little girl would realize that her parents couldn't protect her forever. But, at the very least, parents owed their children a serene childhood surrounded by affection during which "their fine confidence will last as long as possible."

Marie's communications with Meloney also kept pace, and in 1929 Meloney sent Marie a gorgeous address book as a Christmas present, bound in leather and embossed with Marie's initials. In it, Meloney had written all the American addresses of everyone Marie had met during her travels there just a few months before.

A short time later, in early 1930, Marie had her fourth and final operation for cataracts. Afterwards, she seemed pleased with her progress, writing Eve that she was able to enjoy brisk walks along narrow mountain trails without fear of falling—and without wearing her glasses.

Limping more noticeably than ever, Meloney traveled to Paris to be with Marie for her operation. Eve asked to accompany Meloney on a visit to London and Meloney obliged. Upon her return to Paris, Meloney tried to persuade Marie to visit America a third time. This time, Herbert Hoover wanted Marie to tour a few of the places she had missed previously, such as the laboratories of Stanford University and the California Institute of Technology at Pasadena. The trip never materialized and this was to be the final time the two close friends would meet. Meloney had recently gotten over a serious case of peritonitis and was feeling so poorly that Marie wasn't sure she'd ever see her again. There was nothing to indicate that Marie might be the first to go.

A letter to Marie written by Eve in July 1933 from Évian-les-Bains, a lovely mountain city known for its spring waters, intimated that Meloney might be coming for another visit in August.

But that was not the case.

Meloney recovered her health, going on to become the organizer of the *Herald Tribune*'s annual Forum on Current Problems, which gathered together notable personalities to speak on various subjects. She also contributed a chapter to an antiwar book called *Why Wars Must Cease* that was published by Macmillan in 1935. As the years passed, Meloney grew more frail, but she continued to edit the new *This Week* magazine from a suite in the Waldorf-Astoria hotel.

In January 1933, Hoover wrote to Meloney, saying he couldn't end his term as president without expressing his appreciation for her collaboration and loyal support over many years.

"I owe to you many additional obligations of purely personal character," he wrote.

Back in Europe, Marie suffered again the clash between family responsibility and scientific pursuits in 1930 when she had to head to

Geneva for the annual meeting of the League of Nations International Committee on Intellectual Cooperation. When its members admitted that they had been unable to meet their goals because they had expanded too quickly, Marie reminded everyone that she had warned members against spreading themselves too thin. She had also scolded them for spending committee funds to entertain representatives of international organizations. To cut costs, she recommended reducing staff and redistributing funds. The others agreed and appointed her to oversee a reorganization. Between meetings, Einstein worked on his unified field theory equations—part of his theory of relativity—and could often be seen consulting with the white-haired Marie.

One day during this meeting, upon being approached by a *New York Times* reporter, she recalled thinking the first time the two met that Einstein was "the funniest man," someone perpetually obsessed by the problem of relativity. For some reason, Einstein didn't seem to get along very well with the other scientists attending this meeting, besides Marie. A few complained that he was distracted by his anger over the committee's refusal to become more political. And yet, at the tenth anniversary of the League of Nations in 1930, Einstein remarked, "I am rarely enthusiastic about what the League has accomplished, or not accomplished, but I am always thankful that it exists."

But the American press seemed more enamored with Einstein by 1930 than they were with Marie. The *New York Times*, for example, described Marie as behaving at the meeting like a brooding recluse—one not nearly as affable as Einstein. According to the paper, she rudely turned down all social invitations extended after the business meetings—something that wasn't really new. As always, Marie preferred to turn in early rather than have to make mindless small talk with people she barely knew. On the other hand, according to the press, Einstein was a charming attendee with an engaging smile.

Even America's fickle young ladies who had so looked up to Marie during her 1921 visit to their country seemed to have moved on to other women for their inspiration. When the sociologist Lorine Pruette asked 347 teenagers in 1924 to name a hero, only three cited Marie. The more glamorous Cleopatra beat her out thirty times over.

When Marie returned to Paris from her various journeys, Eve typically met her on the platform at the train station. But Eve traveled a lot herself during these years. In 1933, she spent a great deal of time in Cannes, glad to be "far from Paris" and eager to take advantage of sea and sun. In a letter posted April 30, Eve expressed how disturbed she was that her mother wasn't getting enough rest, and she encouraged Marie to take Bronya with her on a scheduled visit to Madrid in May.

Marie wrote to Eve on April 26 that she regretted not seeing Eve upon her return from the Pyrenees, writing "I continue to feel tired." She also wrote about a visit with her grandchildren, noting that little Pierre was walking and that Hélène was as charming as ever. "Her hair, at the front, was bothering her and so, without consulting anyone, she took some scissors and cut if off," she wrote. "That shows you her decisive mind."

As Marie's star began to dim, it became apparent that it was another Curie woman's time to shine. Increasingly, Marie began handing over many of her responsibilities at the laboratory to Irene and to Frédéric, who were on the cusp of setting off on a meteoric rise of their own.

Although they had married in 1926, Irene and Frédéric didn't fully collaborate until a few years later—but collaborate they finally did by conducting a string of important experiments on radioactivity from 1931 to 1934. But success only came following a very rocky start that demoralized them both.

From the outset, the pair should have had quite the advantage over other scientists, what with almost two grams of radium at their disposal from which they isolated 200 millicurie of polonium, the most powerful source of alpha rays. (Irene had worked through her second pregnancy to help Frédéric amass the impressive supply.) With this amount of polonium, greater than the supplies held by any other laboratory in the world, the two certainly had all the material they could possibly have needed to perform investigations related to radioactivity and the nucleus of the atom. They also had the benefit of a Hoffman electrometer, a new model in France and an important purchase for the institute because of its high cost.

Nipping at the couple's heels, though, was a trio of highly competent researchers: Lise Meitner in Berlin, Ernest Rutherford in England, and Niels Bohr in Copenhagen, all of whom were investigating the same area of science.

The impetus for the quest shared by all of them had come from Germany where, in 1930, two scientists named Walther Bothe and Hans Becker had made a bizarre discovery. The two had placed a radioactive substance next to a nonradioactive one and then studied whatever was emitted. In short, they were trying to figure out what would happen when individual particles from the radioactive substance bombarded the nonradioactive substance. The two claimed that a new kind of radiation was given off, one so powerful it could penetrate a layer of lead some ten centimeters thick. Although many explanations were offered for the phenomenon, Irene was never completely satisfied with any of them. Irene, who assumed like most scientists that the atom contained only protons and electrons, enlisted the support of her husband to find out more. The two repeated the experiment and were convinced that the deeply penetrating radiation was actually emanating from a new type of gamma ray, one that moved at the speed of light.

They published their findings in January 1932, the same year as the birth of their son Pierre, and presented them at a gathering of physicists a short time later.

But Rutherford, who had discovered the positively charged particle known as the proton within the atom's nucleus in 1919, doubted the couple's assertion that gamma rays were being emanated. He asked one of his talented young researchers, James Chadwick, to look into the matter. Chadwick borrowed polonium from an American hospital and repeated the experiments. Working constantly while getting only about three hours of sleep a night, he came across and identified particles inside the nucleus with no electrical charge that he named neutrons. He published his findings under the title "Possible Existence of the Neutron" and went on to win the Nobel Prize in Physics for his discovery in 1935.

It seems that Irene and Frédéric had come across the same results but had made mistakes when interpreting their own data—and so they lost out. Chadwick's discovery revolutionized the study of radioactive elements. One can only imagine the disappointment Irene and Frédéric must have felt when they read Chadwick's announcement of his findings in the February issue of *Nature*.

Irene and Frédéric also squandered success when a California physicist named Carl David Anderson barely beat them to the discovery of the positron. Frédéric had been studying neutrons in a Wilson cloud chamber—an airtight environment used for detecting particles. He had set up two cameras to record the movement of particles in tiny droplets of condensation in order to determine a pattern. When he finally saw one, he and Irene were stumped by what looked to be some kind of positive particle the same size as a negatively charged electron. They concluded that the latter had somehow gotten into the cloud chamber and bolted right for the neutrons. Anderson replicated the experiment in the summer of 1932 and reached a different—but correct—conclusion: the cloud chamber had displayed the existence of an electron with a positive charge, and it was one that Anderson named "positron."

When news of the discovery was released, Frédéric was vacationing by himself in L'Arcouest as Irene had felt too ill to travel following a bitterly cold winter. (She attributed her increasing exhaustion only to her tuberculosis—and not to her exposure to radiation.) When Frédéric returned to Paris in September, he re-examined the photographs he'd taken of his experiments and clearly saw the positron for himself. The dismayed couple had missed out on a second chance for a Nobel Prize, which was awarded to Anderson in 1936.

But then, thanks to the couple's determination, the two finally caught a break.

Using their polonium's strong alpha rays, they began to bombard various elements, looking for more positrons. Again, their findings surprised them. It seemed the lighter elements sometimes ejected a neutron followed by Anderson's newfound positron—instead of the expected proton.

They reported their findings of the unexpected ejection of a neutron and a positron at the seventh Solvay Conference in Brussels in October 1933, where about forty of the world's most esteemed physicists, including the likes of Niels Bohr and Paul Langevin, had come together again.

(Indeed, the couple was flattered to be invited.) Instead of Marie being the only woman present—as was usually the case—there were two other women scientists in attendance: Irene and also Lise Meitner. Although this conference's focus was on physics, the couple was presenting to a tough crowd.

But it was Meitner—a brilliant chemist and mathematician who was called the "German Marie Curie" by Einstein—who sparked controversy when she criticized Irene and Frédéric for their findings. Speaking on the couple's behalf, a visibly nervous Frédéric had told the conference that he believed they had made a new discovery. He explained how they had jointly discovered the neutron and the positron—without realizing it—and displayed cloud-chamber photographs as proof. Then he explained how they had, in newer experiments, bombarded medium-weight elements with alpha rays and spotted protons. But then, when they bombarded lighter elements such as boron and aluminum, they sometimes saw protons, neutrons, and positrons. They concluded that in certain cases a neutron and a positron—instead of a proton—could be emitted together in the transmutation. What this indicated, Frédéric asserted, was a proton that was a compound which, in turn, indicated a nucleus that was more complex than previously thought.

But Meitner disputed their findings. The intimidating woman stood and announced that Irene and Frédéric could not possibly have observed neutrons because, in her similar experiments, she had never found a single one. Meitner also assailed the couple for following Marie's old-fashioned methodology. Meitner's opinion was considered sacrosanct, and her division of the Kaiser Wilhelm Institute had been heavily funded, until that point, by the German government. Her goal was to be the first scientist to completely uncover every aspect of the atom. After her pronouncement, Irene and Frédéric sat in stunned silence. Sitting near her daughter was Marie, who also said nothing.

But Meitner had a lot on her mind.

More than a year earlier, on July 31, 1932, an election in Germany had reaped a major victory for Adolf Hitler's National Socialist Party with Hitler becoming chancellor at the end of January 1933. Before long, the government was making the lives of Jewish professionals almost intolerable as it sought to purge them from all government agencies and universities. Meitner, who had converted to Protestantism, initially felt shielded from the anti-Semitism. But in September 1933 she was told that she could no longer teach at the University of Berlin. She also wouldn't be allowed to publish her articles, to lecture, or even to attend scientific conferences in Germany. By 1933, Albert Einstein had renounced his German citizenship and moved with his wife Elsa to Princeton, New Jersey, after telling a friend "I am afraid that the epidemic of hatred and violence is spreading everywhere." Many lamented Europe's loss and the New World's gain. As Meitner arrived at the Solvay Conference in October 1933, she felt that the institute she had built from scratch was being stolen right out from under her. She knew that Marie and her ambitious daughter and son-in-law were able to thrive with no similar constraints.

At the conference, Bohr came to the couple's rescue and encouraged Irene and Frédéric to repeat their experiments despite Meitner's criticisms. "After the session," Frédéric later recalled, "we were quite downhearted, but at that moment Professor Niels Bohr took us aside, my wife and I, to tell us that he found our results very important."

They followed his advice and, in December 1933, began a new round of experiments. They stuck with the idea of bombarding an aluminum target and then withdrawing the polonium with its alpha rays. Frédéric noticed that—although neutrons were no longer ejected—positrons continued to show up. Even when the polonium source was removed completely, the clicking of the Geiger counter continued, meaning it was continuing to record positrons. To test the theory, Frédéric moved the polonium, the source of the radiation, farther and farther away from the target. The same thing happened. An excited Frédéric yelled out for Irene to hurry down to the basement to see for herself what had happened. He repeated the experiment, removing the polonium source of alpha rays and watching as the Geiger counter aimed at the aluminum target continued making noises for several minutes. Had the normally stable aluminum become radioactive, or was the Geiger counter simply off for some reason? You'd think the pair couldn't wait another minute to find out. But unbelievably, they had promised to attend a formal dinner that evening and had no intention of changing their plans.

So Frédéric asked Wolfgang Gentner, the young German physicist spending a year with them, to see if their Geiger counter was on the mark. As instructed, he tested it and left a note for the couple with the results. When Frédéric read it first thing the next morning, he felt a "child's joy." The Geiger counter was working perfectly, which meant that he and Irene had indeed come upon a new and important phenomenon: artificial radioactivity. Alone at the time, Frédéric later recalled how he "began to run and jump around in that vast basement as I thought of the consequences which might follow from the discovery." He raced off to find Irene. The two of them then verified their discovery by using other elements—only to come up with similar results. By bombarding aluminum with alpha particles, they obtained radioactive isotopes of elements that were not normally radioactive.

Their long-time scientist friend Jean Perrin presented the couple's new report at a meeting on January 15, 1934, four days after Frédéric's first observation, in which the pair concluded: "For the first time it has been possible to make certain atomic nuclei radioactive using an external source."

The significance of their finding was multifarious. They discovered that a normal element could become a radioactive one through human intervention. Like Pierre Curie before him, Frédéric quickly grasped the power of this discovery, noting that "scientists, building up or shattering elements at will, will be able to bring about transmutations of an explosive type." In his history of modern physics, Emilio Segre called the discovery of artificial radioactivity "one of the most important . . . of the century." Frédéric couldn't have been happier. As he later said, "With the neutron we were too late; with the positron we were too late—now we are on time."

And Frédéric couldn't have been more right—as he and Irene were finding out. With two small children to care for, their work was far from finished as they still had to provide physical and visual proof of their discovery. But the worst part was that Marie wouldn't be around long enough to see how it all unfolded.

CHAPTER 10

The End of a Quest

For most of her life, Marie had not been political. Scientists, she believed, should steer clear of any kind of activist movement—even the feminist kind. But Irene was a very different animal, a thoughtful woman unafraid of taking a public stand on issues she cared about. Irene respected her mother's noncommittal position but, perhaps influenced by Frédéric, she also didn't hesitate to ask Marie to sign a petition on behalf of two men—Italian immigrants named Nicola Sacco and Bartolomeo Vanzetti—Irene believed had been wrongly accused of murder in the United States. Now a well-traveled woman of the world, Marie agreed. This was only the second petition Marie had ever signed, the first being the one on behalf of her British friend Herta Ayrton's jailed suffragettes many years before.

It's not exactly clear why Irene was so engrossed by this case but both she and Frédéric often became swept up in the battles against perceived social injustices championed by their close friend and Frédéric's mentor, Paul Langevin. By this time, Langevin, who had been a foreign member of the Russian Academy of Sciences since 1924, had become convinced that the U.S.S.R.—not the United States—offered the best hope for the future of mankind. Langevin increasingly equated the struggle against fascism with the struggle against war—which he despised—and he was filled with respect for the Soviet Republic's Marxist socialism.

To Irene and Frédéric—as still to Marie—Langevin's opinions were golden, and when Langevin asked for help fulfilling his long-time goal of setting up a Workers University with Marxist leanings in Paris, both Irene and Frédéric signed on without hesitation, even offering to teach occasionally.

Elected a member of the Academy of Sciences in 1934, Langevin was one of the most influential men in France. And he was never afraid to go against the grain. Despite the growth of several anti-German groups in France, Langevin invited Einstein to accept a chair at the College of France in the early 1930s. Facing death threats from the Nazis, Einstein and his wife, Elsa, decided instead to move to the United States in 1933, where the scientist took a position at the Institute for Advanced Study in Princeton, New Jersey. Later, in 1940, he became a US citizen, as did so many other European scientists.

Having visited Russia in 1928 and 1931, Langevin had concluded though that the Soviet style of science—seemingly dedicated to solving urgent social problems—was the style to adopt. He urged Frédéric to join him in the summer of 1933 after he, along with a handful of other European scientists, was invited by the Russian Academy of Sciences to attend ten days of conferences and lectures in Leningrad and Moscow. Irene, though, was off recuperating in the mountains—her home away from home during the late 1920s and 1930s—and was not able to accompany him. She would later join Frédéric on various trips to Moscow as she became, along with the likes of Pablo Picasso, one of the central figures of the world peace movement formed to fight against the idea of another world war.

Langevin, who would go on to join the Communist Party in 1944, was one of the few scientists not afraid to actively take up causes in an effort to help France, which was undergoing enormous political and economic upheaval in the 1930s. One of the biggest headaches was rising unemployment resulting from the economic crisis of 1929. Indeed, because the United States was the main financier of postwar Europe, the Great Depression fostered financial troubles in countries from Germany to Britain—and to France.

In early 1934, tensions came to a head on the streets of Paris as some forty thousand outraged Parisians, organized by far-right leagues, demonstrated on the Place de la Concorde, near the seat of the French National Assembly, only a mile from the Radium Institute. A major motivation behind this display? A man named Serge Stavisky, who had disappeared with a quarter of a billion francs of the public's money in an investment scam. His case put a spotlight on corrupt relationships between government and business, and the official indifference to Stavisky's many duplicitous dealings drew the public's scorn. In fact, the rioters were incensed not only by the Stavisky scandal but by the whole French system. This anger reached fever pitch after police opened fire and killed fifteen protesters. Langevin and his son André later marched in a counterdemonstration of 120,000 people jointly organized by Communists, Socialists, and unions, partly in an attempt to thwart the rise in fascism that Langevin feared would put a damper on scientific research.

Langevin's political bent influenced Marie and in January 1934 she wrote Eve to express her own extreme displeasure at the Stavisky affair. "I find it deplorable that these men who have sufficient training to understand align themselves so easily with the foolhardy. . . . [I]t's this easy taste for luxury that drives weak people to be imprudent because many amongst them have without a doubt not gained much in terms of illicit revenues or even anything . . . but to accept an invitation to a fashionable restaurant seems to be an irresistible attraction one must believe no matter who is doing the inviting," she wrote.

Even Eve was becoming more political. During a trip to Spain with Marie in 1932, she took to heart the feelings of Spaniards as they explained their support of the Second Spanish Republic. That government had won power in April 1931—after King Alfonso XIII fled—on a platform of creating a more modern Spain by establishing freedom of speech and extending voting rights to women. By the time of Eve's visit, popular sympathy for the monarchy—which had maintained a stranglehold on Spain for nearly fifteen centuries and had become a symbol of oppression—had almost completely evaporated.

"It is very moving," Marie wrote to Irene, "to see what confidence in the future exists among the young and among many of their elders. . . . What interests me a lot are conversations with republicans and the enthusiasm they have for revitalizing their country. I hope they succeed!"

Unlike Marie, Irene had always been determined to use her celebrity status as a bully pulpit on nonscientific matters. And no doubt the discovery of artificial radioactivity had helped both Irene and Frédéric to become even more well known. Around the world, scientists such as Niels Bohr talked of how the pair's work had launched an "entirely new epoch" in nuclear physics. In Berkeley, California, Ernest Lawrence, who had invented the cyclotron—the granddaddy of today's most powerful atom smashers—in 1929, was bowled over when he read about the couple's discovery in *Nature* and *Time* magazines. While researchers like Irene and Frédéric learned much about the structure of the atom during these years, atom smashers later helped scientists learn more about the complexity of the structure. They do so by taking a particle and speeding it up to almost the speed of light, colliding it with an atom, and thereby revealing what's inside.

In February 1934, *Time* magazine singled out both Irene and Frédéric. "Husband and wife work like one person with two heads, four hands, 20 fingers," the magazine said. According to a published quote from Frédéric, "we compare notes and exchange our thoughts so constantly that we honestly don't know which of us is the first to have an original idea. Don't you agree, ma chèrie?"

Although Irene and Frédéric might have made the greatest strides during this time as they helped lay bare the inner workings of the atom, competition was coming at them from every direction, especially from the United States, which boasted more resources than Europe.

Playing catch-up, Lawrence spent an entire weekend in late February 1934 bombarding element after element with deuterons—stable particles composed of a proton and a neutron—with the help of his cyclotron. He observed radioactive isotopes popping up everywhere. No doubt a disappointed Lawrence figured out that his lab could have made the same discovery as Irene and Frédéric at any time during the past six months—but sadly hadn't. "We have been kicking ourselves that we haven't had the sense to notice that the radiations given off do not stop immediately after turning off the bombarding beam," he said.

As when Marie and Pierre discovered natural radioactivity, Irene and Frédéric's discovery of artificial radioactivity also was a boon for the medical field, making it possible to create radioactive materials with less money and in a faster and easier fashion. No longer would doctors have to pay the high costs and endure the laborious work of separating naturally occurring radioactive elements from their ores, a process that had long been a drag on the development of nuclear physics.

But for Irene, who admired her mother above all else, the greatest satisfaction came on January 15, 1934, when she and her husband briefed Marie on what they'd found. Indeed, the pair reproduced the entire experiment for both Marie and Paul Langevin, an indication of just what an integral part Marie's former lover continued to play in the family's life. The re-creation gave Marie concrete proof that the work she and Pierre had started so many decades before would be capably carried on. They even presented Marie with a tube containing the first sample of an artificially created radioactive isotope in the same way Marie and Pierre used to dole out little tubes of their own radium preparations to the scientists they most admired.

Ever the consummate experimenter, Marie wanted to verify the contents herself and so took the tiny vial in her radium-burned fingers and held it right next to a Geiger counter. Frédéric remarked later that he would never forget the expression of joy and incredulity that lingered on her face as she listened to the instrument's telltale clacking.

In Marie's letter to Eve in January 1934—this time it was Eve who was away in the mountains seeking the fresh air the Curies always believed would bring about better health—she relayed how an enthusiastic Irene and Frédéric had just completed some important work on a new radioactive phenomenon they had discovered. But she added that she'd have to explain what the couple was up to later as it was "too long to describe by letter."

On January 29, 1934, Irene and Frédéric followed up their discovery with a second paper to the Academy of Sciences titled "Chemical Separation of the New Elements that Emit Positive Electrons" in which they formally announced that they had created artificial radioelements that acted exactly like natural radioelements.

The lifework of two generations of the Curie family had become complementary, with back-to-back breakthroughs building on one another. The work of Irene and Frédéric had opened up a new understanding of the nucleus of the atom and thus provided an extension of the research performed by Marie and Pierre. The latter two had discovered natural radioactivity, which was the property of a few elements, and their study led to insight into the atom's structure. The next generation took this find a step further by showing how scientists could duplicate this natural phenomenon artificially. Thousands of new radioactive elements not seen before now could be created by artificial radioactivity, a term Frédéric always detested. In Frédéric's view, he and his wife had discovered a radioactivity that was just as natural and important as that discovered by Marie and Pierre.

No matter what they called it, both Irene and Frédéric agreed that witnessing the re-creation of their experiment was "without a doubt" the last great thrill of Marie's life.

Marie delighted in the strides being made by her daughter, particularly now that the triumphal research that had been so much a part of Marie's life for nearly five decades was making a slow but steady exit from her regular routine. In some ways, Irene was so much like her mother, often away from her own two children, back in the laboratory within days of giving birth to both of them. But in other ways Irene showed herself to be much more in tune with the idiosyncrasies of traditional family life even when her children were quite young—perhaps because she had Frédéric there to share them with her. (When Pierre died, Irene was only eight and Eve was not yet two.) During a vacation at the seaside in September 1931, Irene wrote to Marie of a tea party she had given to celebrate Hélène's fourth birthday. She even listed each of the toys Hélène had received as gifts: one fish, one small boat, two little ducks, and a doll that was supposed to be unbreakable. "But it broke the first time it fell on the cement," Irene wrote, noting that the fish, too, had been crushed by one of Hélène's playmates. "Hélène faced this adversity with a spirit of resignation and declared that, luckily, the boat and the duck which had not yet been destroyed were by far the nicest," she went on. "Hopefully she will keep this perspective."

Marie's letters to Irene were often more technical. A month later, in October 1931, Marie wrote from a key conference on nuclear physics being held in Rome to complain about a "lack of clarity" in the speeches being given by some of the biggest names in science. "I have, up until now, very little to tell you except that Bohr strongly insists on the impossibility of currently applying quantum mechanics to the interior of the nucleus," she wrote. Indeed, at that conference Bohr famously charged that it was impossible to offer a detailed explanation of the constitution of the neutron.

Starting in December 1933, illnesses begin to crop up more frequently, gripping Marie more ferociously, and lasting longer before letting go. Just a few weeks before witnessing her daughter's world-famous experiment, X-rays revealed she had a large gallstone. This worried her immensely since her own father had died from complications following a gallstone operation in 1902. Marie made up her mind to refuse an operation. Instead, she decided to dramatically change her diet and, of course, get out into the fresh air more often. The remedy worked and she regained much of her strength. Even so, the thought of dying suddenly bore down on her and she wrote Meloney in early 1934 to make sure again that there were no legal loopholes in the agreement regarding the disposition of the radium given to her by the women of America. As always, she insisted that the radium must not leave her Radium Institute, even after her death. Meloney reassured her friend that all was in order. But what Marie never found out was that Meloney failed to comply when Marie wrote to also ask that all her letters be destroyed. "They are part of me and you know how reserved I am in my feelings," Marie wrote.

Marie no longer focused solely on work but sought comfort in the warm embrace of family and friends. During 1933 and 1934, she probably relaxed and enjoyed herself more than she had in her entire lifetime. During a winter holiday spent with Irene, Frédéric, and her seven-year-old granddaughter in the French Alps, the indomitable older woman gave everyone a good fright by going off without a word, wearing snowshoes, only to return hours later after having spontaneously decided to trek a long way in order to see the sunset over Mont Blanc. A frantic but relieved Irene scolded her mother like a child.

Marie was feeling so much better by February 1934 that she actually entertained the notion of going skiing. The only thing holding her back, she wrote Irene, was a sore wrist. In that same letter, she also wrote of her sadness over the sudden death of the Belgian king, Albert I, a leader she had always admired, who had died in a mountaineering accident while climbing alone in Belgium.

In the spring, Marie also enjoyed some memorable times with her sister Bronya as they gallivanted across the French countryside, with Marie eager to show off its loveliness. From Marie's letters, the trip—spread out over March and April 1934—was quite the adventure. A worried Irene wrote to Marie that she hoped her mother wasn't being "imprudent" as she drove along the zigzagging roads of the mountainous region of the Massif Central before heading off through the Pyrenees and finally on to the fortified city of Carcassonne. The two sisters, of course, were no longer young girls and—as Irene predicted—Marie grew exhausted after driving her small Ford across the country for hours on end.

As they sat one night in a cold and damp room by a quickly built fire in Marie's bougainvillea-covered villa in Cavalaire in Provence, Marie uncharacteristically sobbed in her sister's arms like a sick little girl, according to Eve's biography of her mother. Marie talked of fears over her declining health and how they might stop her from finally finishing the second edition of her two-volume book on radioactivity. She was also determined to build a new house in Sceaux, far from the city. Marie's breakdown was a rare one, and it's possible that a stubborn cold had simply deflated her positive spirits. She was, however, able to recover enough to spend a total of five weeks on the road with her sister, who then returned to Poland. But if nothing else, Bronya felt she was leaving her sister in worse shape than when she had arrived.

Although warned time and again by doctors to rest, an obstinate Marie ignored their advice and insisted on going back to her work at the Radium Institute in April 1934 while also proofreading her book. One day in May, though, she told a colleague at the institute that she was heading home early because she felt a fever coming on. Outside in the garden she called out to a laboratory technician to look after a drooping rose in need of watering. As she got into her waiting car, Marie warned him through the window one last time not to forget. She would never see him—or that rose—again.

Both Eve and Irene realized their mother must have felt seriously ill when she finally agreed to be taken to a clinic. A round of tests failed to provide a definitive diagnosis, however, although doctors were so concerned they recommended she be taken to a sanatorium in the Savoy mountains.

Eve enjoyed feeling useful again as she accompanied Marie there, staying by her bedside for weeks on end. Marie's brother and sisters also helped keep her company on and off through June, with Irene expected to take over during August. By then, it was assumed, Marie would be fit enough to return home to Paris.

But one day Marie's fever suddenly spiked above 104 and doctors told Eve, in confidence, that Marie most likely wasn't going to hang on much longer. Determined to protect her mother from the fact that she was dying, she chose not to call the family to her bedside. She also put a stop to any treatment that would only prolong her pain. But Eve was frightened. Indeed, she wrote to Meloney that she couldn't bear to see her mother suffer and that she often had to leave the room so that Marie wouldn't catch her crying.

When awake, Marie would comment absentmindedly about one experiment or another. On July 3, Marie took her own temperature, noting happily that it had finally returned to normal. After speaking with doctors, though, Eve found out it was only a sign that the end was near. As Marie gazed out through an open window at the sunshine, she must have been feeling like her old self as she clearly asserted that it wasn't the medicine that was making her better, but "the pure air . . . the altitude."

By this time, Irene had arrived, but the normally stoic researcher was too distraught to stay in the room with her mother. Therefore, the duty of keeping watch at her side was left to Eve and the doctors. Marie's last coherent words before she took her last breath on July 4, 1934, were a protest to the doctor about to give her an injection. "I don't want it," she said. "I want to be let alone."

Eve, who was holding her mother's icy hand until the very last minute, would remember that "she was all in white . . . her white hair laying bare the immense forehead, the face at peace, as grave and valiant as a knight in armor, she was, at this moment, the noblest and most beautiful thing on earth."

The doctor said that it was something of a miracle that she had lasted to the age of sixty-seven. He pronounced the cause of death to be "aplastic pernicious anemia. . . . The bone marrow could not react probably because it had been injured by a long accumulation of radiation."

But even in her final weeks Marie never acknowledged that her beloved radium had anything to do with her health problems, insisting instead that all she ever needed to get back on her feet was just a bit more time in the outdoors. She was an avowed agnostic to the end, and there was no priest present and no prayers said during her final days. Irene, who lost not only a mother but also a colleague, was devastated. She later said that she had been profoundly influenced by her mother all her life and that, especially during her childhood, "could never have imagined her having any human weaknesses."

Marie's death made front-page headlines around the world. Interviewed on vacation at Watch Hill, Rhode Island, Albert Einstein spoke of her "ingenuity, extraordinary energy, and unusually independent progress in politics and in social matters." Bohr wrote to Irene that "it must bring great comfort knowing about all the encouragement and pleasure Madame Curie in her last year had from the wonderful discoveries of you and your husband which . . . have crowned her great life work." A more emotional Georges Fournier, one of Marie's students who had worked with her at the institute, dramatically exclaimed that "we have lost everything."

At exactly noon on Friday, July 6, 1934, on a bright sunny day, the coffin of Marie was slowly lowered over Pierre Curie's in the small cemetery in Sceaux, a site she had visited so faithfully over so many years. There were no official speeches or ceremonies, and no important dignitaries were invited to pay their respects. Outside of the family, there were only close friends: the Borels, the Perrins, the Regauds, and, of course, Paul Langevin. The sole grandiose gesture was an amassing of some gorgeous wreaths, including one from the president of Poland. Only a handful of reporters huddled about, but within minutes Frédéric had chased even them all away. Perhaps fearing criticism over the low-key affair, Eve wrote that she and her sister "believed that we were respecting the intimate wishes of our mother, in burying her in the cemetery in Sceaux, in the tomb where Pierre Curie rests, and also in giving the ceremony a simple character."

An Associated Press article on the funeral said that not a word was spoken at the ceremony, noting that the silence was broken only by the constant singing of birds perched in nearby trees. "It was just 10 minutes from the time that the coffin arrived until all had gone," the article said.

Pierre's brother, who had always remained close to his sister-in-law, was too weak to come to the funeral. Also absent was Marie's younger sister Helena, who was vacationing in the mountains with her granddaughter and couldn't be reached in time for her to make the trip. But a grieving Bronya and Jozef had journeyed from Warsaw, bringing with them, unbeknownst to each other, the sendoff they knew would have meant the most to their sister. Over the coffin, they each sprinkled a handful of Polish soil.

Five days later, Eve shipped Marie's silver pocket watch to Meloney, who wasn't able to make the trip across the Atlantic. Eve wrote that the watch, which had once belonged to Pierre, "has no value except that she wore it always and liked it. It was on her table near her bed when she died."

Life at the Radium Institute continued nearly as though nothing had happened, with André Debierne taking over as Marie's successor as Irene was not yet experienced enough for the post. After Marie's death the Radium Institute was renamed the Curie Institute in her honor. Irene seemed to have no problems with the arrangement, writing Meloney that Debierne was being very accommodating as part of a situation that was far better than if she and Frédéric had been placed in charge of the large laboratory.

Irene was in no shape to take over the Radium Institute at this stage anyway. By the early 1930s, she was spending huge stretches of time in the mountains, a separation that must have taken an enormous toll on

Frédéric and their children—and even on her progress at work. No one has ever confirmed whether Frédéric might have taken a mistress during these separations, but he was described by many as a flirt and a ladies' man. And yet, the correspondence between husband and wife always betrayed a warmth and respect for one another.

Irene and Frédéric did not receive a Nobel Prize in 1934 but they did, as everyone expected, receive a wire from Stockholm in November 1935 informing them of their selection. Certainly both were thrilled at this incredible honor although Irene's joy was tempered by the fact that Marie wasn't there to share the experience with her. Irene remembered how oppressive the press had been with her own parents some thirty-two years before, and she warned Frédéric to lie low while avoiding reporters. The next day, as the Radium Institute celebrated the couple's achievement with the traditional garden tea, Irene instinctively replied, when congratulated, "In our family we are accustomed to glory." Coming from someone as straightforward as Irene, her words weren't meant to be boastful but were simply a statement of the obvious.

Irene used the occasion of the Nobel announcement to further plug the causes she believed in. Indeed, in a long interview in *Journal de la Femme* on November 23, 1935, she elaborated: "I am not one of those . . . who thinks that a woman [scientist] . . . can disinterest herself from her role as a woman, either in private or public life. . . . If [the Nobel award has thrust my name, the name of a woman, a little more in the limelight than on other days, I feel it is my duty to affirm certain ideas that I believe useful for all French women. Therefore, I have accepted the presidency of several meetings where the rights of women are discussed."

She said she believed that the government must guarantee equal rights for men and women, and she urged all women to band together in a common struggle to advance their cause.

Again, Irene was from a very different generation of scientists, one that didn't shy away from getting involved in nonscientific causes.

When Irene and Frédéric gave their lectures upon receiving the Nobel Prize for Chemistry in December 1935, both husband and wife stood together at the podium. Irene told the audience that the experiments had been carried out by the two of them, working as a team, with an eye toward discounting any false reports that she might have somehow played only a secondary role in their accomplishments.

But the most poignant moment came when Frédéric looked up from his prepared remarks and said, as he looked straight at Irene, "It was certainly a satisfaction for our late lamented teacher, Marie Curie, to have seen the list of radioactive elements that she had the honor to inaugurate with Pierre Curie so extended." Some were surprised that Irene didn't speak more about her mother in the various interviews she gave at the time of the Nobel announcement. But, as Irene confided in a letter to Meloney, "I do not understand how I could speak of her to journalists."

The Nobel ceremony showed off both Frédéric's charisma and Irene's reserve. During a public reception, Frédéric was forced at one point to excuse himself and go out in search of his wife, who was holed up in a corner reading a book. But Irene had managed to make one concession to the occasion: she wore a striking black evening gown, marking one of the few times in her life she looked truly glamorous. Although most everyone was charmed by Frédéric's panache, Irene managed to retain at least one stalwart fan—her little sister. About this time, Eve was asked to write an article about her famous sibling. In it, she revealed not a hint of jealousy, but instead lavished praise on Irene, asserting that she'd never known her to lie, get angry, or say a nasty word. She admitted that she'd always found Irene's lack of interest in her appearance perplexing. But she expressed real admiration for her sister's intelligence—and the way in which she was able to absorb information—both of which she described as products of a brilliant mind.

Meanwhile, Frédéric displayed his own perceptiveness by nearly predicting the atomic bomb during his Nobel lecture, taking note of the great amount of energy released during transmutation.

He told the crowd: "If we look back at the past and consider the progress made by science at an ever increasing pace, we may feel entitled to believe that researchers, building up or breaking down elements at will, will be able to bring about nuclear reactions of an explosive nature— veritable chain reactions [and] one can imagine the enormous release of useful energy which will take place. But, alas, if all the elements on our planet are contaminated, we can only look forward with apprehension to the consequences of the unleashing of such a cataclysm."

A year before, the great Italian-American physicist Enrico Fermi had very nearly stumbled upon nuclear fission as he conducted experiments that built upon Irene and Frédéric's work. As he and his team looked for radioactive transformations, they used the newly discovered neutron to bombard one element after another. The only obstacle they faced was the thickness of the sheet of foil in which their uranium sample was wrapped, foil that blocked the fission fragments that their instruments would otherwise have picked up.

There would still be plenty of scary times in the near future. At that Nobel ceremony in 1935, Hans Spemann—a German embryologist who won the Nobel Prize in Physiology or Medicine—punctuated the end of his acceptance speech with a gesture as Irene and Frédéric looked on incredulously. That gesture was a Nazi salute.

CHAPTER 11

Tributes and New Causes

For many months following Marie's subdued funeral in Sceaux, lavish tributes to her flowed from America.

Nearly a year after Marie's death, in June 1935, Missy Meloney presided over one such marquee gathering on New York City's East Side when Mayor Fiorello LaGuardia officially renamed Exterior Street in the Bronx as Marie Curie Avenue. Some five thousand people turned out for the event along the East River, with the mayor eulogizing Marie as not only a great scientist but also "a great hero of peace." Through the years, countless streets and sites across the United States have been named for Marie Curie, including the lovely Marie Curie Playground that opened in 1956 at 46th Avenue and 211st Street in Bayside, Queens, and was later expanded under the supervision of Robert Moses to become the much-used Marie Curie Park.

Meloney had just been elected first vice president of the prestigious New York Newspaper Women's Club in May 1935 and was exceedingly busy, as always. In addition to keeping close tabs on Irene and Eve, Meloney was a friend and confidante of many movers and shakers, including Eleanor Roosevelt. In one of the First Lady's "My Day" columns, Roosevelt wrote about one particular visit with Meloney—who had suffered from cancer—in New York City.

"Here is a woman who, in spite of months of illness, has managed to keep her guiding hand on the production of a weekly magazine, has given her thought to the arrangements of one of the best known forums in the country, has worked on a book and talked to innumerable people. Her spirit has remained an outgoing spirit in spite of all the limitations of pain and weakness. There is something very stimulating in talking with this gallant woman," she wrote.

On November 23, a few months after the LaGuardia event, the Roerich Museum in New York City sponsored a Curie Memorial Celebration with Einstein making the night's most memorable speech, summing up Marie's scientific achievements in only thirty words but spilling out hundreds more to describe her strength of character.

He spoke passionately of their twenty years of friendship and of "her strength, her purity of will, her austerity toward herself, her objectivity, her incorruptible judgment—all these were of the kind seldom found in a single individual." He went on to add that, "if but a small part of Mme Curie's strength of character and devotion were alive in Europe's intellectuals, Europe would face a brighter future."

Another remark by Einstein, made just before her death and often repeated, noted that "Marie Curie is, of all celebrated beings, the only one whom fame has not corrupted."

Many American students and scientists mourned her death. Their words fed into the image popularized over the years by Meloney. Dr. Joseph Colt Bloodgood, a cancer expert at Johns Hopkins University Medical School, described an encounter with Marie. "Two years ago when I was in Paris attending a scientific meeting of the French Association Against Cancer I brought to Mme Curie a poem from an American actress whose life had been saved by radium," he said. "Countless lives are being saved and the ravages of cancer are being controlled by the use of X-ray and radium." The physicist Robert Millikan, president of the California Institute of Technology, also issued a public statement: "In spite of her continuous absorption in her scientific work, she has devoted much time to the cause of peace. . . . She embodied in her person all the simpler, homelier and yet most perfect virtues of womanhood."

Obituaries in American newspapers sang Marie's praises; the *New York Times* concluded that "Few persons contributed more to the general welfare of mankind and to the advancement of science than the modest, self-effacing woman whom the world knew as Mme Curie."

A year after her death, Marie's last book about her work was finally published. Edited by Irene, it delivered a closing message to young students of the sciences everywhere. At Marie's Radium Institute, where work had continued with scarcely any interruption, the enormous volume was placed alongside the many other scientific tomes available in the bright library. On the gray cover was the name of the author: "Mme Pierre Curie, Professor at the Sorbonne. Nobel Prize in Physics. Nobel Prize in Chemistry." As an emotional Eve later put it, the title of her mother's last book was made up of just one severe—and yet radiant—word: "Radioactivity." Over the years, Marie's many written works have been translated from French into English and continue—to this day—to be sold to new generations of aspiring scientists.

Marie's granddaughter, Hélène, said she doesn't remember much about Marie or her childhood, noting with a laugh that her parents never told her she had such a famous grandmother until she was a young woman.

"But my parents did speak a lot about my grandmother when I was growing up, so much so that it's hard for me to separate memories of when I was actually there from the stories they told me," she said. "I remember being with my grandmother in the Luxembourg Garden in Paris and I know she loved gardening a lot."

Whether Marie's legacy was responsible or not, it's notable that she died just as sciences such as physics were expanding with new fields such as microbiology and biochemistry emerging.

National Research Council data reveal that the number of women earning doctorates in the sciences in the United States climbed impressively from the time of Marie's first trip to America in 1921 to the years just after her death in 1934—from about 50 per year in the early 1920s to about 165 per year in the late 1930s. Yet because the number of men earning doctorates in science shot up even more than did that of women in the 1930s, the women's percentage of the total declined after 1932 from a high of 15.5 percent in 1920 to about 11.5 percent in the late 1930s. It's not clear why this drop happened although figures show a rising number of women earning doctorates in education during this period, perhaps indicating the better job prospects in schools at the time.

Under Marie's guidance, from 1919 until her death in 1934, the chemists and physicists at her Radium Institute published a remarkable 483 works, including 31 scholarly papers and books written by a very prolific Marie. During the same period, the biomedical section of the institute was also fruitful, treating more than eight thousand patients, usually successfully.

But discrimination persisted. Women working in the sciences generally earned less than men and were placed in fewer decision-making positions both in the United States and in Europe. Yet massive changes were in store for Irene and Eve and Frédéric—and for their old friend Paul Langevin—in almost equal measure.

When Irene and Frédéric won their Nobel Prize, the press fawned over Frédéric much more than it did over Irene—just as it had paid much more attention to Pierre some thirty-two years earlier. But an avalanche of accolades poured in for their research. Irene was very much in tune with the discrimination her mother had experienced throughout her lifetime and, even at a young age, felt the dissatisfied voice of an ardent feminist stirring inside her. Marie hadn't been as bothered as her daughter by such matters. Already back in 1925, when Irene obtained her doctorate in physics, Irene bluntly told a *New York Times* reporter during a brief interview that she believed "that men's and women's scientific aptitudes are exactly the same."

Both Irene and her husband were very much a team during their work on artificial radioactivity. As Hélène succinctly put it, her parents worked between 1930 and 1935—side by side—to discover artificial radioactivity, "showing that certain nuclear reactions produce radioactive elements that are unknown in nature."

Their Nobel monetary prize of about $41,000 allowed the couple to pay for their new home at Parc de Sceaux, near Paris. Hélène was eight years old in 1935 and says she can't recall hearing her parents talk much about their winning of a Nobel Prize because they were all so busy during this time shifting from Paris to their new house in the suburbs. Already a good student when it came to math and science, Hélène interestingly adds that, when she was a girl, she remembers that her mother, in particular, gave her the impression that one didn't have to be "a genius to be a scientific researcher . . . otherwise I wouldn't have pursued this myself."

Hélène, who went on to become a respected nuclear physicist and the wife of Paul Langevin's grandson, said her parents never pushed science on her or on her brother, Pierre. But she said they were very much involved in their children's studies and, when separated, Irene and Frédéric always mailed math problems to Hélène and Pierre—just as Marie had always mailed math problems to Irene.

Because of the variety of new opportunities that presented themselves, Irene and Frédéric decided to work separately after winning the Nobel Prize, although they consulted with one another on almost a daily basis.

In short order, Frédéric began rising through the ranks, promoted at Paul Langevin's urging to the position of director of research at the Caisse Nationale des Sciences—or the National Fund for Scientific Research—the highest rung on the career ladder there. Showing off his powers of persuasion, his first course of action was to convince that prestigious body to buy and transform an old electrical plant in Ivry, an industrial suburb southeast of Paris, into what would become known as the Atomic Synthesis Laboratory.

With the help of a three-million-volt impulse generator, the laboratory would soon roar into action, providing man-made radioactive elements for researchers who needed them, saving them both time and money. In 1937, Frédéric was also offered a chair in nuclear physics and chemistry at the College of France. There, he continued his study of the possibility of nuclear chain reactions and nuclear energy production. He was keenly interested in exploring every application of artificially radioactive isotope possible, especially in the area of biology. His Russian aide, Lew Kowarski, who had begun working with Frédéric in 1934, described him as "the most ambitious man since Richard Wagner . . .

who wanted to be Beethoven . . . Shakespeare and Caesar all rolled into one," a person who wanted nothing less than to "revolutionize biology." No doubt Frédéric was enjoying this stage of his life, just before the pressures of war were about to hit. For the first time, he had a modern laboratory, a full professorship, generous funding, and a talented staff at his beck and call. No matter what other appointments he might hold in the years ahead, Frédéric would remain Ivry's active director until his death.

The only thing he didn't have was Irene right by his side. They were working only a short walk away from one another, but it wasn't quite the same as being in the same room all day, with their heads stuck together over a single experiment.

Even so, their careers continued to be intimately intertwined. And Frédéric relied on Irene in every way. As Hélène put it, her father ran almost everything he did, work-wise, by her mother, who became a professor at the Sorbonne in 1937. During this time, Irene's research on artificial radioelement forms in uranium had brought her incredibly close to the discovery of nuclear fission—or the splitting of an atomic nucleus that results in the release of a vast amount of energy. In 1938, the German scientists Otto Hahn, Lise Meitner, and Fritz Strassmann officially became the first to recognize that the uranium atom actually split when it was bombarded by neutrons, although Meitner often isn't credited with this finding. Later, news of the splitting of the atom and its overwhelming possibilities was delivered by Niels Bohr to scientists in the United States and ultimately resulted in the Manhattan Project. Although Irene and Frédéric, in their respective roles, controlled nearly every aspect of nuclear work in France, Hélène recalls hearing them say regretfully in 1938 that, if they had only worked together, they could have been the ones to discover fission.

Irene and her husband also enjoyed a mutual passion for politics, and they both became increasingly politically active after Marie's death, spurred on by the lead-up to World War II.

The rise of Hitler had prompted the French Communists to change tactics and, in 1935, all the parties of the Left and the Center had combined to form the Popular Front. That party, supported by both Irene and Frédéric, won a sweeping victory in June 1936.

The party's socialist leader, Léon Blum, immediately selected three women with progressive views to participate in his government; they included Irene, who served as undersecretary of state for scientific research from June to September 1936. She decided to accept what had always been designed to be a temporary post created just for her, according to Hélène, in an effort to affirm the influence of women. Even in that short time, Irene was able to help lay the foundations, with scientist Jean Perrin, for what would later become the National Center for Scientific Research, a massive institution employing some 25,000 people and covering all fields of knowledge. "Scientific research, [my mother] always said, is a comforting field from the moral point of view for the pleasure of discovery, even if it has little practical importance, for the sentiment that all new knowledge is definitively acquired for humanity," Hélène said.

As Irene wrote to Meloney, "Fred and I thought I must accept [the post] as a sacrifice for the feminist cause in France, although it annoyed us very much." Unlike in Marie's lifetime, "disinterestedness" was no longer the aspiration of Irene's generation and, instead, "responsibility" was the new mantra. But, because it forced her to be diplomatic, the government position was never the job for her. One day, according to biographer Rosalynd Pflaum, a new secretary brought for her signature a letter declining to attend a government function. That secretary believed Irene had inadvertently omitted the customary phrase "I regret I cannot attend" and so had added it herself. Irene refused to sign until this phrase was removed because she wasn't sorry in the least that she was not able to attend the function.

In the United States, Meloney also was becoming more political. As early as May 1933, she had joined a radio symposium on "Literary Freedom and Nationalism." That symposium, and subsequent forums, attacked the Nazi party for burning books. In May 1933, the Nazis had decreed that any book "which acts subversively on our future or strikes at the root of German thought, the German home, and the driving forces of our people" was to be burned. Asserting that bigotry had no place in a nation of intelligent people, Meloney had said at the radio symposium that the only weapon with which to fight was "the courage to establish before the civilized world a standard of right thinking and right living and then persistently to support that standard."

Irene, still suffering from tuberculosis, wasn't feeling well that summer of 1936 and was forced by the end of July to seek treatment at a resort in the mountains. With Irene due to accompany him to Moscow in the fall, Frédéric was only too happy to have his very thin wife away if it meant she might regain her health and also put on some weight. "I want to have a beautiful Irene with me," he wrote. She wrote him back with a warning, as she often did: "Don't smoke too much if you want to be my beloved." She was gone long enough for Frédéric to admit, "I miss you in spite of your terrible . . . bossiness. Perhaps that is what I miss." Irene knew her husband was most likely enjoying himself and wrote back, "I am glad to know that you are getting a good rest, probably by going out at 4 A.M. to check the fishing nets and going to bed at midnight after having danced, not to mention having played tennis in between."

Upon returning to Paris, she met with Blum and happily turned over her government post to Jean Perrin, a close family friend who had been awarded the Nobel Prize in Physics in 1926. Then, in September, Irene and Frédéric went to Moscow to attend the first Mendeleev Conference, where Frédéric was asked to give the opening address, on the discovery of artificial radioactivity, which had given new meaning to the Mendeleev periodic classification of the elements. The couple was wined and dined in Moscow while being introduced to many important people. In total, they were away about three weeks while little Hélène and Pierre stayed with a nanny.

By the summer of 1937, excavation began on a thirty-foot-deep cellar at Ivry to house Frédéric's new pet project: a particle accelerator—better known as a cyclotron—that was fast becoming the best weapon yet in the physicist's battle to make more discoveries. To expedite its installation, Frédéric had sent an aide to Berkeley, California, for a year's internship with Ernest Lawrence, who was by this time building his fourth cyclotron on the University of California campus. Lawrence had won a patent for his invention in 1934 and would go on to win a Nobel Prize for the cyclotron in 1939. Early in 1935, Frédéric—like Lawrence—had turned to the Rockefeller Foundation in the United States for financial help in building his cyclotron—and the foundation obliged. Indeed, eager to promote the promised medical applications of neutron therapy, the Rockefeller Foundation helped fund the construction of nine cyclotrons between 1935 and 1945. Three were in Europe with one proposed by Bohr, one proposed by Frédéric, and one by the University of Stockholm. The other six were in the United States.

When Irene returned from Moscow, she launched into a series of experiments that would occupy her for more than two years and that would result in her most outstanding individual work, with some believing she was on the verge of winning a second Nobel Prize.

Irene was making strides with the help of a new teammate, Paul Savitch, a smart young Yugoslav who got along well with the blunt Frenchwoman. The pair had been studying uranium when, during one particular experiment, they thought they had observed the release of a neutron and a new radioactive isotope. This new isotope, with a half-life of 3.5 hours, displayed properties similar to lanthanum—a silvery white metal lighter than uranium.

But others such as Lise Meitner and Otto Hahn—still critical of Irene and Frédéric—said it wouldn't be possible for an element lighter than uranium to come out of the reaction. They said they had conducted similar experiments and gotten different results. Meitner was so convinced that Irene was mistaken that she shot off a severe warning to her. Unless Irene publicly retracted her findings, Meitner would have no choice but to publish a renunciation of her work. At home, Irene and Frédéric discussed the disagreement ad nauseam, and Frédéric couldn't help but bring up the issue with Hahn himself when they bumped into each other at the Tenth International Congress of Chemistry in Rome in 1938. Frédéric defended Irene's results, but Hahn, considered the greatest radiochemist of his time, stood his ground. He belittled Irene's findings as inaccurate, arguing that they were the result of antiquated testing methods. By then the Germans had made a sport of poking fun at the French, referring to the new 3.5-hour substance as "Curiosum," meaning it just had to be a derivative of Curie. In actuality, Irene and Savitch were within a hair's breadth of discovering nuclear fission and didn't realize it. They thought they had simply discovered a new element.

On May 1, 1938, Irene and Savitch published a report making the same claim, but one that was less confident in tone than their first report. Even so, they asserted that "this substance cannot be anything except a transuranic element, possessing very different properties from those of other known transuranics."

Thanks to their combined efforts, Irene and Frédéric had been among the first to understand how a chain reaction could be set off in a mass of uranium. Indeed, it wasn't long before Frédéric—who had predicted the release of nuclear energy through a chain reaction in his 1935 Nobel Prize speech—had actually sketched out the basics of a nuclear reactor. To this day many scientists argue that if war hadn't broken out, Irene and Frédéric would have been the first to discover an atomic pile.

Soon enough, Hahn and his partner Meitner had much more pressing issues to worry about. Indeed, Meitner, considered a Jew, was by the late 1930s more concerned with her very survival than with disproving any of Irene's theories. Hahn, too, was wrapped up in trying to help Meitner escape permanently to Sweden. While the exile from Germany ultimately saved her life, it also cost her the ability to publish with her colleagues, which denied her a Nobel Prize as well as a prominent role in many books on the history of physics.

Irene and Frédéric, who both espoused a desire for peace—just as had Marie and Pierre—truly believed that only good would come from their discovery of artificial radioactivity. France, after all, needed energy, a need that could be filled by nuclear power. But neither could have imagined the repercussions of their research.

Throughout their time together, Marie and Pierre had opted to publish everything they discovered. This was also the attitude adopted by Irene and Frédéric for the discovery of artificial radioactive isotopes. But angst resulting from the tinderbox that was Nazi Germany—combined with a growing awareness of the dangers that could crop up if the Nazis ever got their hands on information surrounding chain reactions—led them to stop publicizing their work. With Hitler becoming more aggressive by the day, on October 30, 1939, they placed all their documentation on nuclear fission—including the principles behind nuclear reactors—in a sealed envelope, which they deposited in a safe at the Academy of Sciences, where it would remain secret until 1949.

Incensed by France's policy of appeasing the dictators, Irene and Frédéric, along with Jean Perrin, crafted an open letter to French Prime Minister Édouard Daladier that read in part: "Our external interests are being entrusted to very weak men. We demand that no concession be made to the Italian and German demands." Both Irene and Frédéric were even more outraged when Daladier—a supporter of appeasement—then joined with Hitler and Mussolini to sign the Munich Agreement on September 29, 1938, an agreement that allowed the Nazis to annex areas along Czech borders.

In agreement with Irene and Frédéric was Winston Churchill, who was among the politicians who berated his own government, with Neville Chamberlain at the helm, for behaving dishonorably by signing the Munich Agreement. Churchill proclaimed that "the partition of Czechoslovakia under pressure from England and France amounts to the complete surrender of the Western Democracies to the Nazi threat of force. Such a collapse will bring peace or security neither to England nor to France."

To further protest France's appeasement policy toward Hitler and Mussolini, Frédéric and Paul Langevin helped put together a delegation that landed a meeting with French President Albert Lebrun, even though he exercised little real power as president, according to biographer Denis Brian. At that meeting, the first members of the delegation to speak reportedly went off on emotional tangents as Lebrun placated them all by listening patiently and not interrupting. When it came to his turn, Frédéric deliberately spoke in a calm and cool manner—but he was the one to ruffle feathers nonetheless. "Has not the minister for foreign affairs shown more than compliance?" Frédéric asked. "Are we not faced with a complicity verging on treason?" At the mention of the word "treason," Lebrun expressed outrage, jumping up from his seat and angrily ordering the entire delegation out of the room.

A short time later, Frédéric gave a speech at the Sorbonne on the medical benefits of artificially radioactive isotopes. In the audience sat

Lebrun. At the end, he took Frédéric aside to tell him how much he had enjoyed his talk. And then he surprised Frédéric by suddenly apologizing for reacting so strongly in his meeting with Frédéric's delegation. "In the presence of so many people I could not have acted otherwise," Lebrun said. "But you were right!"

With the Communists becoming the only members of the French Chamber of Deputies to formally protest the Munich Agreement—and with the Socialists losing credibility as a party that was strong enough to ward off the fascist dictators—it's no surprise that Irene, Frédéric, Langevin, and other progressive intellectuals increasingly threw their support behind the Communist Party.

An indignant Irene wrote of her outrage in a letter to Meloney: "That traitor . . . Chamberlain: France and the United States are not much better. Sometimes I think that this civilization will really disappear, for fascism is going back directly to barbary, and the democracies are entirely in the hands of the private interests of capitalism."

On September 3, 1939, France declared war on Germany, five hours after Britain's declaration, and that evening Paris experienced its first blackout. Frédéric was drafted into the army and everything was about to change. Irene and Eve were horrified by Germany's invasion of Poland at the end of September and were worried especially about all the relatives they had still living there. The day after the declaration of war, Irene wrote to twelve-year-old Hélène, who remained with Pierre at L'Arcouest under the care of a nanny, surely bringing back memories of the nearly identical correspondence her own mother had sent her when Irene was staying at L'Arcouest at the beginning of World War I.

According to biographer Denis Brian, Irene wrote, "What is happening is sad, but you have heard us speak often enough of these matters to know that we are not surprised. It is possible that we may be separated for some time. Paris is very calm and it is hard to imagine that there is a war. I have seen . . . groups of children . . . being evacuated. Pinned on their clothes is a large ticket on which is printed their destination. . . . [They look] like small packages waiting to be shipped." She also reminded her daughter to study her music and to practice her gymnastics daily, and to make sure her little brother went into the water often while she kept watch on the shore.

Irene, who continued to work at the Curie Institute, wrote to Meloney of her contempt for the Nazis. But she was certain they would never dare touch a member of the Curie family and she vowed to follow in the footsteps of her mother. She knew Marie would never have abandoned her radium or her laboratory under any circumstances. She also told Meloney how Frédéric had been put in charge of a group of laboratories to organize their work for war purposes.

According to biographer Denis Brian, she wrote, "I will also work in that direction. . . . We hope very much the U.S. will abrogate the neutrality act. This law is unjust and is contrary to the interests of the U.S. for, if there is finally a fascist victory, it is rather clear to see how fascism would establish itself, first, in the countries of South America, getting to the North afterwards. If the U.S. does not help the democracies, at least by selling arms, I think it will be a crime against our common ideal of civilization. . . . [I]t will also be the greatest political fault."

The goal of getting the United States more involved in the war was shared by both Irene and Frédéric. But it was actually another Curie who had the greater impact when it came to this issue.

CHAPTER 12

All about Eve

To the scientific tasks at hand, Irene and Frédéric gave due attention, working steadfastly in a way that would have made Marie proud. Eve's situation was more tenuous. Although she didn't share the scientific passions of her mother or her sister, the poised beauty had been Marie's nursemaid, her confidante—a mainstay during the elder woman's last year. When Marie was worried about the state of her brother-in-law's health, she wrote Eve—not Irene—to unload her sadness and concern over his old age and deafness. After Marie's death, Eve no doubt felt a bit unsure about what to do next. Musically inclined, she had performed her first piano recital in 1925 to encouraging reviews, practicing long hours on a gorgeous grand piano given to her by her mother. Sometimes, Eve's amiable brother-in-law even played along with her, as an amused Irene and Marie looked on.

But music was not a career Eve would be able to bank on. She entered into it a bit too late in life, and she was smart enough to know it. After Marie's death, Eve corresponded regularly with Meloney, whose judgment she trusted implicitly. Eve's role model helped nurture the pursuit of a second career choice—writing. Meloney encouraged her to do some translating and to put her knowledge of the arts to good use by becoming a music and movie critic for a number of magazines. Eve hadn't wanted to trade on her family name, but she wanted to earn enough money to take care of herself—and she'd always enjoyed a way with words. And so she began writing articles on theater, music, and movies under a pseudonym for various Parisian periodicals. As early as 1932 she had translated and adapted the American play *Spread Eagle*, written by George S. Brooks and Walter B. Lister, for stage production in France. (The play had premiered on Broadway in April 1927.) Eve's French version premiered in Paris in October 1932 and enjoyed a long run under the name *145 Wall Street*.

But soon enough, Eve found an even more ideal calling. Although her charms were often eclipsed by the scientific achievements of her immediate family members, the artistic Curie was remarkable in that she was able to develop her distinct talents in a field foreign to that in which she had been brought up. And, in the end, her life turned out to be perhaps the most adventurous one of all.

Not long after Marie's death in 1934, US publishers approached Eve with the idea of writing a biography of her mother. Initially, she was terrified to take on such a monumental task, but she also didn't want the job to go to someone who might not be as kind or as truthful. And so she accepted the offer, writing Meloney to say that she felt up to the challenge even if there was still so much she didn't know about her own mother's life. She worked out of a small apartment in Auteuil, a borough of Paris that was once home to Victor Hugo, collecting and sorting through stacks of documents and letters left behind by her mother. She also journeyed to Poland during the fall of 1935 to enlist the help of Marie's family and childhood friends—no doubt a therapeutic exercise for a twenty-nine-year-old woman who was still mourning her mother.

As Eve noted in the introduction to her biography, Marie was already thirty-seven years old when Eve was born and, by the time Eve was mature enough to really know her, Marie had become a famous physicist revered around the world—one often away from home. As a child, Marie had been a fleeting presence. It wasn't until Eve was a grown woman, and Irene was married, that she was able to get a sense of the "real" person behind the scientist, and even then some of the details were sketchy.

Although Eve kept putting it off, and had to be constantly badgered by her editors at Doubleday to finish it, she ultimately did justice to her mother's history when her biography *Madame Curie* was finally published in 1937. In weeks, the book became one of the best-selling biographies ever written. Vincent Sheean, an American writer and reporter based in Europe, was enlisted to translate Eve's original French version into English, and the biography was published simultaneously in England, France, Spain, the United States, and elsewhere.

In the United States, *Madame Curie* was an instant smash that garnered a long list of honors. In 1937, for example, Eve received a National Book Award from the American Booksellers Association. The book was the December 1937 choice of the Literary Guild and was chosen by the American Library Association as "the best non-fiction book of the year." It also won for Eve the Clement Cleveland Medal awarded annually by the New York Cancer Committee for the year's outstanding achievement in the fight against cancer. Proof that Eve's work was an authoritative one is the fact that there was no other biography of Marie until Robert Reid's 1974 *Marie Curie,* which played up the implications of research on radioactivity during the World War II and Cold War eras.

Eve's biography was written with a daughter's sentimentality with nary a mention of the specifics related to the scandal involving Paul Langevin in 1911. (Years later, the book was often criticized for this omission.) She also praised Irene throughout the book as an endearing and highly intelligent personality. At the time the *New Yorker* called the biography a rare book, "one which reconciles us to belonging to the human race." Eleanor Roosevelt enthused, "I have read it with great thrill. The simplicity and beauty of the style and the understanding and love for her mother are in themselves wonderful. Another review simply said that "This is a great book because its subject is a great subject."

The success of Eve's biography quickly piqued Hollywood's interest, and Universal Studios immediately bought the film rights with the comically skilled Irene Dunne in mind to play Marie. Dunne even traveled to Europe to meet with Eve to discuss the project, although nothing ever came from that meeting. A few years later Universal sold the property to MGM as a vehicle for their popular Swedish star Greta Garbo. Writers such as Aldous Huxley (whose script was dismissed for being too literary) and F. Scott Fitzgerald took stabs at adapting the screenplay for

Garbo. Altogether, eighteen writers had a hand in the final product. One of them, Salka Viertel, Garbo's close friend, spent time in Paris doing research for the script. According to "A Rose for Mrs. Miniver: The Life of Greer Garson" by Michael Troyan, Viertel wrote to the film's director Sidney Franklin afterwards that "Irene Joliot-Curie dislikes, or rather, is very much hurt by the idea that the life of her mother should be shown in a motion picture. Confidentially, I found out from other people in Paris that Eve never told anybody that she had sold her book to the movies. Irene and her husband have great influence and are very much respected here in France. As she is not certain what our picture will be like, even if the intentions of my collaborators are the best, she refuses any cooperation, since she would thereby lose the right to protest. . . . And, dear Sidney, when she refuses it is as if the Rock of Gibraltar were to refuse."

Irene couldn't see Garbo as her mother in the movie, and the idea was shelved when Garbo left MGM in 1941 and the country was thrown into World War II. Finally, though, production got back on track, with the Irish American beauty Greer Garson—an actress more to Irene and Eve's liking—cast as Marie in a script credited to Paul Osborne and Paul H. Rameau. She won the part over such acclaimed actresses as Joan Crawford. The role of Pierre Curie was initially supposed to go to Spencer Tracy, who thought it the perfect opportunity. But because production took so long to get off the ground, Tracy was forced to move on to another project. The part eventually went to Walter Pidgeon, a wise choice since he ultimately won an Academy Award nomination for his performance. Pidgeon later remarked that one of his all-time favorite scenes was when Pierre goes out to get a nice piece of jewelry for his wife and is killed. Of course, that never happened. Pierre was killed by a horse-drawn carriage—that part was true—but he wasn't on his way to buy a present for Marie. Interestingly, Van Johnson, at the time a young song-and-dance man, was chosen rather than a female actor to play the role of the reporter who landed an interview with Marie—supposedly Meloney.

Although the producers romanticized the Curies' life, as might be expected, the sets and costumes were authentic thanks to meticulous weeks of research. Stacks of old family photos, supplied by Eve, were pored over. A physicist named Dr. Rudolph Langer was even hired as a technical assistant for the movie production to make certain the "science" parts looked real but were also easy for a layperson to follow. One of the directors, Mervyn LeRoy, later said in his autobiography that he absolutely refused to direct any scene unless he himself completely understood what was going on in it. One of Langer's main assignments was to recreate, in detail, the experiments conducted by Marie and Pierre for the screenwriters. However, he found his biggest challenge was to convince MGM set designers not to make the laboratory in the movie look like something out of the gothic melodrama *Frankenstein*.

Later in her life, Greer Garson talked often of the challenge of working in Garbo's shadow. She said in interviews that during the many months of shooting Madame Curie had become her obsession. She was intrigued by the similarities between the two of them. For example, one of Garson's favorite vacation spots as a young girl had been L'Arcouest in Brittany where, of course, the Curies had always vacationed several times a year.

As Garson studied Marie's background, she came across a poem—apparently unknown even to Marie's daughters—that had been one of Marie's favorites. It was called "To The Young" by Adam Asnyk and, when the screenwriters struggled to come up with a suitable conclusion for the movie, it was Garson's idea to have Marie read that poem at the event marking the twenty-fifth anniversary of the discovery of radium in 1923. As admiring scientists looked on in the scene, Marie [Garson] told them: "Leave, then, the dreams of yesterday. You—take the torch of knowledge and build the palace of the future."

Garson often said that Marie Curie had consumed her thoughts and imagination in a way no other character had ever consumed her before.

Despite the fact that the final product scarcely mentioned Marie's sister Bronya and other important family members, Eve wrote a letter of approval to the producers saying how much she liked their film adaptation of her book. The movie was honored with seven Academy Award nominations including Best Picture, Best Actor, and Best Actress. Ultimately, however, it took home none, losing out to *Casablanca* for Best Picture in 1943.

Most important for Eve, perhaps, was that her book allowed her to take a months-long journey across the United States in early 1940 (the first of many interactions that would ultimately lead her to reside in America permanently) that brought the most unfamiliar Curie to the attention of Americans as an outspoken personality in her own right.

Indeed, nearly twenty years after she had made her first trip to America alongside her mother and Irene in 1921, Eve arrived in the United States in January 1940 aboard the Italian liner *Vulcania* for a lecture tour in English that was ostensibly about promoting the biography of her mother but was actually, for Eve, all about getting Americans involved in the war effort.

Following the September 1939 invasion of Poland, where Eve still had many close relatives—and with the Nazis occupying France in 1940—nothing was more important to a patriotic Eve than stopping Hitler.

Tall, slender, with fair skin and dark hair, Eve cut a glamorous figure and was considered by many to be one of the most beautiful women in Paris in the 1920s and 1930s. In a heavily publicized shoot admired by people around the world, Eve had been photographed for French *Vogue* in December 1937 wearing a dress by the famous Italian fashion designer Elsa Schiaparelli—Coco Chanel's greatest rival—in addition to a black felt hat and, most notably, long black antelope gloves.

It's not surprising that Eve's arrival in the United States—where her smiling face graced the cover of *Time* magazine in February 1940—created a stir not unlike the one sparked by the Curie women's trip in 1921.

Eve's thirty-city tour was supposed to have happened a year earlier but—because of her extensive duties as the recently named director of women's wartimes activities for the French Ministry of Information—the trip had been postponed until early in 1940. No matter where she went in America, she carried her political message with her. (Her essay "French Women and the War" was published in the *Atlantic Monthly* in early 1940, and many of her lectures were given under the same title.) Just after arriving in Manhattan, for example, she paid a surprise call on Mayor Fiorello LaGuardia at City Hall, looking elegant as always in a simple black silk dress and matching hat. She set the tone for the entire tour when she told a reporter on the scene that "France is proud to have your sympathy but there is a feeling that America is needed in this conflict."

Later Eve gave a lecture, again titled "French Women and the War," at the non-profit Town Hall founded by suffragists in 1921 during which she complained that French women had no political rights—but had been bearing much of the war's burdens. At midday, she was guest of honor at a Hotel Astor luncheon hosted by Town Hall trustees and designed to address a different topic—"The Arts and America." But, again, Eve's message was squarely focused on politics at the gala attended by some 1,500 arty guests including actor Paul Robeson and author Dorothy Thompson.

Without specifically referring to Hitler, Stalin, or Mussolini, she said that France had been obliged to enter the war in order to defend spiritual values. "The most terrible condemnation of regimes of oppression which today rule several countries of Europe is the sudden and total disappearance of their art," she said.

On January 22, the Philadelphia Smith Club sponsored a lecture by Eve at the Academy of Music in Philadelphia. For a change, this time the focus was more on Eve's mother, who had received an honorary degree from Smith College in 1921. A year's worth of planning and selling tickets to the event culminated in a huge money-maker for the Philadelphia Smith Club, with Eve delivering a speech on "The Magic of Radium" before a massive audience.

On February 2, Eleanor Roosevelt invited Eve to have dinner and stay the night at the White House. The First Lady mentioned the occasion in her syndicated newspaper column "My Day." She wrote that, "the President was very glad to see her again. . . . I looked at this slender, dark, very chic and charming woman, who does not look as though she were made for hard work, and yet can come over to this country and spend two months on the road. She looks her best on all occasions, meets people, I am sure, with the thought in her mind that she is not only making friends for herself, but for her country and that, therefore, she must try to meet as many people as possible to draw out their questions and their points of view and, if possible, leave them with a friendlier feeling toward her nation than they had before."

In a funny exchange, Eve told reporters in Washington after being asked about her choice in careers that "I don't hate science . . . it just terrifies me."

Following in her scientific mother's footsteps, though, Eve made several stops at American universities.

In April, she traveled to Harvard, where she spoke to an overflowing crowd at the New Lecture Hall.

She explained how French women had enthusiastically taken on new positions in society, fitting fuses for cannon shells and performing every other kind of job imaginable. She stressed the reasons that made war inevitable to the Allied peoples and painted a vivid picture of everyday life and changes in a nation getting hammered on all sides. She spoke about how her country's schools and universities had bravely continued educating French students as though no war had ever started.

"French women are carrying on, supplanting the men at the front in every capacity," she said to rousing applause.

Articulate and elegant, she explained that, as of January 1940, not only were women, including housewives, demonstrating against the Germans, but some 300,000 women were also working in French munitions factories.

She hammered away at the same subject to a crowd at Stanford University, again in impeccable English, emphasizing the sacrifices of French women and their determination to get rid of the Hitler regime once and for all.

But perhaps her most moving speech was to the American Booksellers Association in New York City on April 9, just as the lecture tour was coming to a close.

She said: "We discovered that peace at any price is no peace at all. We discovered that life at any price has no value whatever; that life is nothing without the privileges, the prides, the rights, the joys that make it worth living and also worth giving. And we also discovered that there is something more hideous, more atrocious than war or than death; and that is to live in fear."

The *New York Herald Tribune* lauded the speech in an editorial: "It is not always easy to discover the reasons for a great speech. There must be an audience, able and ready to respond. There must be a speaker at once inspired and inspiring. To these must be added an extra spark . . . capable of touching off the mysterious power that is eloquence. Some such combination of events clearly surrounded Eve Curie. . . . Those who attended the luncheon . . . in the shadow of the news from Norway . . . are still talking of her stirring presence and [the] poignancy of her voice."

The editorial was referring to news of Germany's invasion of Norway early in April. Yet, much to Eve's dismay, Americans remained divided over whether to get involved or stay out.

Most people don't realize the extent to which Eve influenced French politicians and the role she played in the war effort. Just before the fall of France in June 1940, Prime Minister Paul Reynaud was in a quandary after finding out that Britain could not send a substantial number of aircraft to France. Because of this, he was told by his vice premier, Henri-Philippe Pétain, that there was nothing left for him to do but to make peace, adding that "if you do not want to do it, you can hand over the government to me." Reynaud refused, deciding instead to turn to President Franklin D. Roosevelt for reassurance that America would come to the rescue if need be. In this effort, Reynaud received help not only from US Ambassador William Bullitt, but also from Eve. Archives show that a first draft of a letter from Reynaud to Roosevelt actually was composed mostly by Eve. Later, in November 1942, Eve pleaded the French government's case herself to Roosevelt: "Your voice can be a guide . . . a rallying point. You can show them the way."

Eve left the United States late in the spring of 1940, having made a startling impression on Americans. Later, in a broadcast of "Betty Crocker Magazine of the Air," Eve talked about what she had discovered about the country during her trip: "The European idea used to be that America was a nation of practical people all wrapped up in making money. I discovered that nothing could be more untrue. This is the most sentimental and idealistic country I know. It's the only country where one may find a genuine freshness of feeling . . . an immediate reaction to world events and sincere indignation over wrongs committed. I have a deep love for America."

It's not surprising that Eve and Irene began parting ways during this period, with Eve aligning herself much more with America and Irene and Frédéric growing increasingly attached to the Communist Party.

Eve was only in Paris a short time before she left again, on June 11, 1940, for England. This time, she was seeking to escape following France's surrender, while her sister and brother-in-law decided they would stay in the country, in large part to protect their stash of radium. Eve took off with 1,300 other refugees aboard a cargo ship used to accommodating no more than 180, a ship that was strafed by German aircraft as it left the French coast at Bordeaux. Three days later, on June 14, Nazis marched into Paris, leaving the French stunned at how quickly their country had buckled under the weight of the German assault. Most of the residents watched and did nothing as swastikas went up on Parisian boulevards—but not Eve. She was always on the move. In England, she joined the Free French Forces of General Charles de Gaulle and started her active fight against Nazism, which resulted in the Vichy government's depriving her of her French nationality and confiscating her Paris apartment in 1941, just as they did with all the "undesirables." Her belongings, including much of Marie's furniture, were auctioned off, a move that Eve later said had broken her heart. Eve spent most of the war years based either in Britain, where she often met with Winston Churchill, or in the United States, where she gave lectures rallying the public to stay behind US involvement in the war.

Basically, Eve grew to be nothing less than the spokesperson for the women of France during World War II, with her most well-recorded quote of the period being "Peace will not come soon, and it will not come at all while the Hitler regime remains in Germany because the French are determined that when this war ends there will be no more fighting in Europe for a long time." Away from her family, Eve found her footing and was poised to assume the mantle of the Allies' conscience. She bravely endured the nightly bombings on London that began September 7, 1940—and also mingled with important people. On September 30, Lady Diana Cooper, wife of the future British ambassador to France, Duff Cooper, wrote in her diary that she had just dined in London with a variety of influential French people including Eve. "The guns, bombs, orchestra and jabber deafen and bewilder. . . . Eve Curie told me that she had been on a tour of provincial ARP [air raid] shelters with Lady Reading [the wife of the Viceroy of India Lord Reading]. All the little children of five have Mickey Mouse gas-masks."

While Eve was in England, Frédéric stayed on in Paris, writing often to his mother. In one letter, dated July 14, 1940, he told her that his two children were safe in L'Arcouest and that Irene's health was improving with medical treatment at a sanatorium in Switzerland. Their work also was safe, locked away in a vault, where it stayed throughout the war. Although tuberculosis was getting the better of Irene, she sneaked across the border often to visit her children and eventually managed to smuggle them back with her to Switzerland. An optimistic-sounding Frédéric wrote Irene on August 9, 1940, that all was quiet in Paris, noting that the city actually had become more livable now that millions had left. But, like so many others, Frédéric had been dismayed over France's defeat following a mere six weeks of resistance and, by the spring of 1942, he had secretly joined the French Communist Party, a major anti-Nazi force at the time.

Frédéric's good friend and mentor Paul Langevin was enduring a seriously rough patch, having been arrested in early November 1940, allegedly for seeking to prevent Germany from getting rid of the Jews. Indeed, the Gestapo ransacked the office of the sixty-eight-year-old revered scientist, hauling him off to prison as though he were a common criminal. Offers of asylum came from the United States and elsewhere, but the Germans rejected every single one of them. Following demonstrations by Frédéric and others, Langevin was finally released after thirty-eight days in jail but was immediately driven to the small town of Troyes, more than one hundred miles southeast of Paris, a place that had no scientific institutions. There he was kept under house arrest for most of the war.

Meanwhile, in January 1941, Eve returned to America aboard the SS *Excambion* to give another lecture tour designed to boost support for the cause of freedom. In 1941, she had published *They Speak for a Nation*, a touching book of letters from French people around the world, often addressed directly to the people of America. At the time, Eve said the book was designed to give the American people an honest portrayal of what the people of France were going through. In one letter, written by a French teacher in Brittany in 1940 to a friend in America, it said: "What a friend this America is. . . ." All profits and royalties from the sale of this book, in which French people thank America again and again for its assistance, were applied to the relief of French prisoners of war.

Again, as during her trip in 1940, she was invited to the White House. But during this lecture tour, Eve encountered some angry opposition to her views from a group of isolationists known as the America First Committee, a group that included the aviator Charles Lindbergh. At the time, the organization was the most powerful isolationist group in the country, with 850,000 members. But it was no matter. Eve had the ear of the president and the support of editorialists at influential newspapers such as the *New York Times*. A few months after Eve returned to London, she joined forces with Meloney, who was still in New York, to try to figure out a way to get Irene and Frédéric out of France. But Irene refused to uproot herself from Paris, where she knew she had the best chance of continuing with at least some of her research. Even when her health worsened again, and Irene was compelled to return to the Swiss mountains, she worked on an overview of the discovery of both naturally occurring and artificially produced radioactive elements. She also put on some much-needed weight.

While Irene and Frédéric, in some ways, led a life very similar to the one they had led before the war, Eve's world was about to be shaken up in extraordinary ways.

By late 1941, Eve's life resembled a child's game in which one gambles by spinning an arrow that lands on a word. In Eve's case, the word was "opportunity." And what an opportunity it was! Because of her talent, and perhaps even more because of her family ties and connections, Eve was asked to become a "Special War Correspondent" on a world tour sponsored by Allied Newspapers Limited in London and the Herald Tribune Syndicate in New York. No one had to ask her twice.

Thanks to Roosevelt, the United States had been sending airplanes, weapons, tools, and food to foreign countries whose security was deemed vital to America's safety under the lend-lease program. Eve snagged a seat on one of the first lend-lease transatlantic flights set to take place on a huge Clipper seaplane. On November 10, 1941, she left the Pan American Airways base at LaGuardia Field in New York City with a fur-lined coat on her lap, on the adventure of a lifetime. Air travel in

1941 was, of course, a vastly different experience from what it is today. According to biographer Richard F. Mould, Eve described her preparation for the trip:

I had not gone to bed that night [before] and had spent my time kneeling on the floor of my Manhattan hotel room, between a scale, meant to weigh my luggage, and my two traveling bags, made of soft canvas, from which I eliminated one by one the heaviest items. The 44 pounds allowed were only 35 pounds once my typewriter was deducted. The 35 became 29 after I had taken into account my papers, stationery, and the thick Anglo-French dictionary with which I could not part . . . 29 pounds! In the old days I had sometimes taken more than that to go on a weekend in Wiltshire or on Long Island. Now the 29 pounds had to do for several months, for the heat of Africa and India as well as for the Moscow winter. . . . A taxi had taken me to La Guardia airport around 4 A.M. with my typewriter and my two bags. I had put my heaviest clothes on, from sweaters to sheep-lined boots, so that the scales of Pan America Airways should be indulgent with my luggage. I had kept on my arm (a classical trick of all Clipper passengers) the three coats that would take care of all weathers. . . . I was leaving for the equator and dressed for Alaska.

The Pan American Clippers were the largest aircraft to be built before jumbo jets; Eve's, called the Capetown Clipper, was due to travel from the United States to the west coast of Africa via Brazil. She noted that the plane was filled mostly with military and airline personnel and that she had "the status of a clandestine passenger." When the Clipper was about to come down on November 13, the steward began wildly spraying the passengers with anti-mosquito "flit so that everyone would be insect-proof by the time of landing in British Gambia."

Eve was about to embark on one of the most professionally stimulating periods of her life, successfully working as a war correspondent who would land interviews with the Shah of Iran, Mohammad Reza Pahlavi; the leader of Free China, Chiang Kai-shek; and Mahatma Gandhi, among others. No matter what else she was doing, though, she always took time to meet with French and Polish soldiers. Eve's reports from her journeys were published in American newspapers and were gathered together in the book *Journey among Warriors,* published by Doubleday in 1943. The best-selling account of her 40,000-mile trip across a series of wartime fronts—North Africa, Iraq, Iran, Russia, India, Burma, and China—was nominated for the Pulitzer Prize for Correspondence in 1944, eventually losing out to Ernest Taylor Pyle. In the book, she spoke in broad strokes about America's role in the world. "I knew that many [Americans] believed, not without bitterness, that we simply expected America to play Santa Claus eternally and that we jealously coveted her riches," she wrote. "Certainly it was a great and frightening responsibility for the United States to be wealthier and happier than anybody else."

Eve was the only member of the immediate Curie family to leave France, and the next few years of her life were to be not only challenging, but incredibly fulfilling and exhilarating.

CHAPTER 13

The Ravages of Another World War

When World War II hit, the Curie sisters went in two directions and rarely saw each other during this time. As a war correspondent, Eve traveled forty thousand miles to wartime fronts that included Iran, Iraq, India, China, Burma, and North Africa. Meanwhile, Irene stayed in France, where by this time she had achieved nearly as much fame as her mother.

Irene's daughter Hélène confirmed that her mother and aunt had almost no direct contact during the war but described "indirect contacts" that usually occurred during those times when Irene was convalescing from tuberculosis in the Swiss mountains.

Although Irene and Frédéric continued their research, at least to some extent, life in France was anything but routine. After the country formally surrendered on June 25, 1940, France was split into two zones. The Nazis directly controlled three-fifths of the nation including Paris—the occupied zone. The rest, unoccupied France in the south, was governed by a French government at Vichy. All citizens, even if they were "celebrities" like Irene and Frédéric, had to use ration stamps and wait in long lines if they wanted to make any sort of purchase. Adapting to this changing environment took a new kind of ingenuity. The normally very straightforward Irene was suddenly forced to be cagey about her every movement. Authorities granted her permission to travel outside Paris only after she fabricated a story about how she'd forgotten some very important laboratory equipment during her stay in the tiny sanatorium town of Clairvivre, France. The reality was that Irene was on a mission to gather up what was left of the radium that Marie had brought back from the United States in 1921. No matter what, Irene knew she had to get her hands on it before anybody else did. Somehow, she was finally allowed to travel to Bordeaux, where the radium was stored in a vault, and two of her technicians were allowed to go with her after Frédéric was denied a permit to leave Occupied France.

Irene was more fortunate than others, though, in that she could at least tap into the interest from the balance of the Marie Curie Fund left over from her mother's first trip to America in order to make ends meet. Thanks to Meloney, who still acted as a sort of fairy godmother to the girls, that money continued to be deposited regularly in a Geneva bank account and proved invaluable in helping Irene and Frédéric cover living expenses during the war years. Others weren't so lucky. During the winter of 1941–1942, for example, there was such a shortage of food that some people were forced to eat cat meat.

Eve may not have always been in touch with her sister, but she was frequently in contact with Meloney. The two wrote letters back and forth, expressing worry over the situation faced by Irene, Frédéric, and their two children in France. But Irene and Frédéric still weren't interested in going anywhere else.

Irene wrote Meloney: "We think we have a mission to fill: to prevent the dispersal of the scientific workers from our laboratories and the loss of the radio-elements needed for our work that were collected by my mother and ourselves. If we find that, for some reason or other, we cannot be useful any more, then we will try [to leave]."

Not long after, Irene reiterated the couple's feelings to Meloney yet again: "Fred is happy to know that it would perhaps be possible, by way of exchange [for another couple] . . . to go to the United States. We both think that we must not do it now, but if the circumstances would compel us to change our mind, we will arrange [to let you know]."

Irene also wrote to Meloney of her hope that the United States would abrogate the neutrality act. "This law is both unjust and is against the interest of the United States for, if there is a fascist victory, it is clear how fascism would establish itself, first in South American countries and then moving to the North. If the United States does not help the democracies, at least by selling arms, it will be a crime against our common civilization."

Although the United States had not yet entered the war in 1940, the American ambassador was extremely active in France. Probably with the encouragement of Meloney and Eve, he made overtures to Irene and Frédéric, offering them the ability to relocate to America. The ambassador had already helped the Curie family's good friend, scientist Jean Perrin, flee to the United States. But Frédéric was adamant that he couldn't leave his research behind and that he would never abandon his fellow scientists with no means of escape.

In 1939, Lise Meitner and Otto Frisch had announced the theory of nuclear fission. But at the same time, in 1939, Frédéric and his team also had demonstrated the possibility of splitting the uranium nucleus. Frédéric was at the top of his game, having been named a captain in the French army. He was asked to lead a group of scientific researchers focused on chain reactions—and in particular the finding of the precise amount of energy released in chain reactions. In addition, the team looked into the requirements for the creation of an atomic pile using uranium and heavy water. Their findings resulted in five patents being taken out in 1939 and 1940 and made Frédéric even more famous than he had been. The Associated Press reported that Frédéric was seeking "to find a way to make a two-dollar pound of uranium produce enough energy as was obtained by burning $10,000 worth of coal."

Again by this time he was perceptive enough—and had enough experience—to recognize that his work was a precursor to a chain reaction that could ultimately lead to the development of an atomic bomb. Because Frédéric had gathered conclusive evidence of nuclear fission and had photos of the fragments to prove it, France very well should have been way ahead of everyone else when it came to developing nuclear energy and an atomic bomb.

But, soon enough, Frédéric was forced to walk a very thin tightrope.

By the time Paris fell to the Germans in June 1940, Frédéric had already had the foresight to smuggle most of his working documents and materials to England with the help of his assistants Lew Kowarski and Hans von Halban. In another astute move, Frédéric had convinced the French at the beginning of the war to purchase the entire stock of heavy water manufactured by the Norsk Hydro Company in Norway. According to Marie's biographer Barbara Goldsmith, the Germans had previously tried to purchase this substance for their own weapons program. Not wanting that to happen, Frédéric arranged to smuggle Norway's supply of 185 kilograms of heavy water by plane to Paris. When the Germans caught wind of this, they headed to the Curie Institute. But by this time, thanks to Frédéric and his team, the heavy water was already on its way aboard containers to England where it was safeguarded—at one point at Windsor Castle—and out of the reach of the Germans.

As a result, Frédéric was still able to work clandestinely with relative success to impact the expansion of fission projects in Britain and America even during the years of the war. The transfer of Frédéric's materials out of France helped to end the Nazis' own nuclear research efforts. But it also meant that the important work was going on in England and, in the end, the French were given only some rights to the exploitation of certain British patents.

Frédéric also convinced the Nazis, despite serious pressure from at least some Germans, that it wasn't necessary to remove the radium supply or even the country's only cyclotron—or atom smasher—from France. It would be easier for everyone, Frédéric contended, if he could keep his equipment right where it had always been. This isn't entirely surprising as the Nazis sought to put French scientists to work for the German war effort as effectively as possible.

According to Hélène, her father was an efficacious operator, a shrewd negotiator who finagled an agreement with the German government that, as the longtime director, he absolutely had to maintain the right to access any part of the lab he wanted, even those parts where a German physicist installed there in September 1940 might be working.

Given the war, though, the Curies decided to halt their work on nuclear fission and to change course—at least to all outward appearances—toward biological and medical applications of their research. Frédéric and his colleague Antoine Lacassagne, for example, used a radioactive iodine tracer to show how well a thyroid gland was working. Irene, Frédéric, and others also realized that if chain reactions could be tamed, fission could lead to a promising new source of power.

Despite tensions with the Germans, all these accomplishments spelled more accolades for Frédéric. In 1943, he was asked to become a member of the Academy of Medicine. That same year he was elected to the French Academy of Sciences. Irene was denied a seat, as her mother had been, because of her gender.

The struggle by the French Resistance against the occupying forces began to consume more and more of Frédéric's time. In June 1941, he had taken part in the founding of the National Front, a French Resistance group founded by French communists aimed at uniting all anti-Nazi elements under one roof. He soon was named its president.

The war rarely reached a point as low for the Allies as it did the following year, in the spring of 1942, when German armies pushed deeper into Russia. The war, and especially the genocide of the Jewish people, really hit home when Irene and Frédéric's colleague, a theoretical physicist by the name of Jacques Solomon, was executed. Frédéric formally joined the outlawed French Communist Party in 1942. (In 1956 he became a member of the party's central committee.) Some who knew the couple say Irene may have been more sympathetic to Communism than was her husband, but she never became an official member, perhaps fearing the impact on her children's future. Through it all, they both continued to denounce the treatment and imprisonment of colleagues caught for working with the Resistance.

Pretending to be busy with mostly medical research, Frédéric also risked his life by using his lab to manufacture explosives and radio equipment for the Resistance, turning it into a clandestine weapons arsenal under the nose of a German physicist and his assistants as the Resistance worked toward the liberation of Paris. Outwardly, Frédéric had little choice but to pretend to cooperate with the Germans, unless he wanted his entire lab confiscated.

Later in 1942, the war took yet another turn for the worse when the Allies invaded North Africa, leading the Germans and Italians to immediately occupy the remaining free part of France. By November 1942, the Germans occupied all of France except for the southern coast, which they gave to the Italians.

Meanwhile, Paul Langevin was still living under house arrest in Troyes, with authorities monitoring every communication relayed to him. Langevin turned into an international symbol for intellectual re-sisters, and scientists around the world, including Einstein in America, lobbied hard to have him set free. But their efforts were to no avail. From time to time, Frédéric sneaked in to visit his mentor—putting himself at risk every time.

With Irene's health worsening, Frédéric had no choice but to apply for permission for his wife to go again to Switzerland over Christmas 1942. The government agreed. Frédéric appeared to be more despondent than usual when he wrote Irene that "It is the first time since our marriage that we have not been together for this holiday. . . . I had a lot of black thoughts, and so did Hélène. But we tried not to show them." He also informed Irene that their son, Pierre, was doing poorly in school, prob-ably due to sheer carelessness. Irene wrote back that "I see that I certainly have an 'Ideal Husband' [the title of a play by Oscar Wilde]. . . . He takes care of the finances, as well as Pierre's homework, and he does not forget his wife. His wife does not forget him either." Irene struck a positive note in her letters to her children, most likely remembering how lonely she used to get as a young girl whenever Marie was away, especially at the start of World War I.

According to biographer Rosalynd Pflaum, the letters exchanged between Irene and Frédéric exude a playfulness and warmth, despite the dire circumstances under which they were written. In the spring of 1943, Irene mailed Frédéric a picture of herself to prove she was putting on some weight. Frédéric wrote back: "I authorize you to gain some more. I love plump women (One will suffice.) Your [round] cheeks fill us with admiration and I am going to be jealous because mine are pitiful."

Fortunately, a few months later, Irene was able to return to Paris again, just as the film version of Eve's biography of their mother was being released. Unfortunately, American films were banned in France during the war.

Eve was in London at this time. Often she spoke in radio broadcasts to the people of France from England, urging them to keep up the fight against the Nazis.

Like the discoveries of her mother and sister, Eve's travels as a foreign correspondent for both American and British publications during World War II became the stuff of legends.

In 1942, when Eve was in Rangoon, Burma's capital, she churned out stories on her typewriter by night while witnessing by day how a large town engulfed in fear can quickly go to pieces just as an individual can have a nervous breakdown. She wrote about how a shop that she visited one minute was suddenly barricaded without warning thirty minutes later. Then the restaurant at which she was to meet some American correspondents locked its doors without warning, for lack of food. She told of an encounter with a brash pilot from the American Volunteer Group who had volunteered to destroy Japanese planes and kill Japanese men long before Japan had attacked the United States. She described him as a type of American she had never encountered before. He told her that "you don't need a uniform to fight the Japs. You need a plane, gasoline, and ammunition. So far I have brought down three Japanese machines— which means fifteen hundred bucks that I will send home. We get $500 per plane shot down, in addition to our monthly pay of $600." He was, Eve said, a fighter without being a soldier, one who loathed anything resembling military discipline.

She then traveled to China, which had been at war with Japan since the summer of 1937. Eve interviewed the well-educated and charismatic Communist general Chou En-lai during a scene she described as something out of a mystery movie. Eve told of being taken to a dingy hideout in the middle of nowhere, one she would never have been able to find again on her own. Inside was a table covered with white linen and decorated with bowls of fruits and flowers. During an elegant dinner, Chou En-lai clarified for Eve the attitude of the Communists towards the political and military leader Chiang Kai-shek. He said that the Communists opposed the reactionary rule of the Kuomintang, or the National People's Party, while nevertheless considering that the Generalissimo was an irreplaceable war leader, the present symbol of Chinese unity. He told Eve that

China must hold out until the Western powers could attack Japan. Eve also met the beautiful Madame Sun Yat-sen. Her husband, Dr. Sun Yat-sen, the revered founder of the first Chinese Republic, had died in 1925. Eve found this woman, still a prominent figure in Chinese politics, living in simple surroundings. Eve wrote that she reminded her very much of Marie in the way that she could be entirely inconspicuous and yet unforgettable at the same time.

From China, Eve went on to Calcutta, where she was fascinated by the impact of British domination over a country with a population many times the size of the occupying force. She admired the fact that Britain had built canals, created large modern cities, and constructed a highly efficient railway system, the fourth largest in the world. But she was appalled that Britain hadn't done more to alleviate the groundswell of unemployment and illiteracy.

Going by train from Calcutta to Delhi, Eve also stopped off to visit Jawaharlal Nehru and his daughter Indira Gandhi in March 1942. At the time of Eve's visit, the future prime minister of India was getting ready for her wedding, scheduled for the very next day. Nehru explained to Eve his views on India's independence, which was becoming a more and more pressing issue in the British Empire. Eve was charmed by Nehru's wit and the fact that he didn't travel with a bodyguard. She urged him to do a lecture tour in America, but teased him that Americans would not be able to understand his strong British accent.

From Nehru, she got a letter of introduction to Mohandas Gandhi. The seventy-two-year-old activist, an icon in his own time, agreed to walk with Eve the following morning at 7 A.M. She remembered the admiration her mother had expressed for Gandhi, how she instinctively shared his belief in a return to a simpler way of life, one that she had not thought irreconcilable with the progress of modern science. He told Eve that India "can win her laurels only through non-violence. What we have achieved in the past 20 years shows that immense results could be obtained if the principle of non-violence were generally practiced by all our people." On no point could Eve find Gandhi in contradiction with his intransigent creed of nonviolence. Therefore, she concluded that Gandhi could never play a part in the government of India during the war, as pacifism, she knew, would only make the Allies lose the war.

Eve also spent a great deal of time in the Soviet Union in 1942, traveling by plane from Tehran along with a British colonel and various other military officials. She began to understand, at least to a minor degree, why the Soviet government held such a fascination for Irene and Frédéric. She was impressed by the sacrifices made by ordinary soldiers and civilians in warding off the invaders. She also said she discovered that freedom of religion was more of a reality in the Soviet Union than anyone in the United States or in Britain knew it to be. She was stunned that almost everyone she met had read her biography of her mother. Indeed, she had had no idea that the book had even been translated into Russian. Eve certainly never sympathized completely with Communist ideology, but she wasn't the system's most rabid critic either.

By late spring in 1942, Eve returned to America. She had left a United States with isolationist inclinations only five months earlier and returned to a country fully entrenched in the war. She had arrived aboard a flying boat traveling via the Near East, Africa, and South America—exactly the same route that she had followed before. In the five months she'd been away, America had rushed headlong into war, forced to move fast by the attack on Pearl Harbor on December 7, 1941. Eve's trip to America was her fifth there since 1921. She wrote that the journey back always filled her with the same swell of emotion. To her, America would never stop humming the words of Walt Whitman: "Liberty! let others despair of you!—I never despair of you."

She returned from her extensive travels with more than enough material for what would eventually be released as a now nearly forgotten 501-page book titled *Journey among Warriors,* a book dedicated to her mother, Marie Skłodowska Curie, "whose birthplace in Poland and whose grave in France lay in lands occupied by the Germans as I made this journey among the soldiers of our war."

After a stop in New York, she visited Washington, as she usually did, and enjoyed another leisurely dinner with the Roosevelts at the White House. Eleanor Roosevelt wrote, "One looks at this chic, well-groomed, delicate French woman and marvels at the calm with which she must have faced many dangerous moments, and one is proud of women!"

In December 1943, General Dwight Eisenhower was appointed Supreme Allied Commander in Europe. By then Eve was back in London as a member of General Charles de Gaulle's Fighting French Women's Corps—the Voluntaires Françaises. Interviewed by Marjorie Avery for the *New York Times Magazine,* Eve said: "When the time comes for us to go back to France we will go with great humility, almost asking the pardon of those who have remained. We have eaten while they starved. We have been warm while they shivered. We have been free to act while they have kept their courage high under domination and military occupation." Eve also remarked to Avery that one must give of one's mind, but also of one's hands and one's bodily strength. "It is the lesson I learned from my mother," she said.

Sadly, another great sorrow had befallen the Curie sisters by this time when Missy Meloney died in New York on June 23, 1943, following a month-long bout of influenza. Just before her death *Time* magazine had appropriately described her in an article on the media business as "fine lace made of cable wire." Just before her death, she had handed over control of her syndicated Sunday magazine section *This Week*—which by this time garnered 18 million readers—to an ex-publicity man named William Nichols.

After a funeral procession from the residence of her son at 7 Washington Square North in New York City, a requiem Mass was recited for her on June 25 in St. Patrick's Cathedral. Not only was former President Herbert Hoover a pallbearer, but so were Owen D. Young, former head of General Electric, and David Sarnoff, president of the Radio Corporation of America. Following the service, attended by more than five hundred mourners, she was buried in Woodlawn Cemetery in New York. Through her countless articles, Meloney had worked to better education, improve health services, and assist a war-torn Europe. Eleanor Roosevelt dedicated her entire column to Meloney on June 29, 1943. It read: "One never came away depressed from seeing 'Missy' Meloney. . . . [I]f I am sometimes weary and think that perhaps there is no use in fighting for things in which I believe against overwhelming opposition, the thought of what she would say, will keep me from being a slacker. She believed that women had an important part to play in the future. She not only helped such women as Marie Curie, who were great women, but she helped many little people like myself to feel that we had a contribution and an obligation to try to grow."

Like the Germans, Parisians knew by 1944 that the Allies, including the United States, would try to invade France in order to liberate Europe. Rumors swirled, and life in Paris plummeted to a new low, with numerous regulations making the occupation increasingly intolerable. A shortage of gas meant that few cars were on the road while a curfew prevented people from being out at night. Coal, and even food, were hard to come by. Worried that the Nazis might use Paul Langevin as a hostage if an invasion did occur, Frédéric and a few other scientists concocted a chancy escape plan in 1944. They purchased forged ID papers and even an armed guard to help get him out of the country. Indeed, after staging a car accident, Frédéric and the other men literally lifted up a bandaged Langevin and dragged him across the border into neutral Switzerland. By this time, Langevin was in deep despair and might not have cared what happened to him. Langevin's son-in-law had been executed while his daughter had been seized by the Gestapo for collaborating with the Resistance and transported by cattle car to Auschwitz. She survived, however, and eventually returned to Paris after being freed by the Soviet army when they liberated the camp at the end of the war. In addition, two of Paul Langevin's grandsons—sixteen-year-old Michel Langevin and his cousin Bernard—were sent to prison for three months for distributing subversive material. Michel Langevin eventually married Irene and Frédéric's daughter, Hélène.

The move to help Langevin escape was a bold one that led Frédéric to harbor doubts about his own safety going forward, especially after he was arrested and released only after an unnerving round of questioning. Certainly, he believed that Paris had become too precarious a locale for his family, so he helped them flee to Switzerland, where they joined up with Langevin. (First, though, an education-minded Irene made sure that their daughter, by then seventeen, completed her baccalaureate exams.) Irene and the two children donned backpacks and hiked across the French Alps into Switzerland on June 6, 1944. It was a fortuitous date on which to do so. With the Normandy invasion going on, no one was paying too much attention to the border crossing. Frédéric went underground in Paris, assuming the name Jean-Pierre Gaumont, and managed the production of even more explosives at the College of France; they were later used during the liberation of Paris. In a farcical turn of events, the college's administrator felt compelled to tell Frédéric's assistants that their boss must have escaped to Switzerland after he failed to show up at the college for a few days. It was a well-intentioned lie, one meant to protect Frédéric. In actuality, though, Frédéric had been hiding out at the home of one of his assistants who enjoyed a good laugh over the administrator's communication.

Just three months after the Normandy landings of June 1944, and around the time of the liberation of Paris in August 1944, a healthier Irene was able to receive clearance to leave Switzerland and return to Paris. Her months of relaxation had done her good, and she returned a new woman to her research at the Curie Institute. Hélène and her brother, Pierre, too, were able to return to Paris without all the formality of applying for official papers. At long last, the family was united again.

On April 12, 1945, President Franklin Roosevelt died of a cerebral hemorrhage, and his successor, Harry Truman, was informed for the first time of the imminent existence of the atomic bomb. On April 30, with the Soviet army about to arrest him in Berlin, Adolf Hitler put a gun in his mouth and committed suicide, just days before Germany's unconditional surrender. Frédéric, Irene, and their children were at L'Arcouest for the first time since the occupation when, on August 6, 1945, they heard that a single bomb, codenamed "Little Boy," had been dropped by the Boeing B-29 bomber Enola Gay, demolishing the Japanese city of Hiroshima and killing an estimated 200,000 people. On August 9, a plane dropped a plutonium bomb on Nagasaki, killing another 140,000. The two attacks forced Japan to surrender on August 14. Irene said often in later years that she was glad her mother had not lived to see the day when atom bombs were dropped on Hiroshima and Nagasaki.

Both Irene and Frédéric considered the use of atomic fission for a bomb to be a betrayal of all they had worked for. Indeed, they both felt partly to blame for all that had transpired. What they had hoped was that fission would lead to a promising new source of power. But instead it had led to massive death and devastation.

Frédéric tried to assuage his sense of guilt by writing in an article published in August: "The immense reserves of energy contained in the uranium devices can also be liberated slowly enough to be used practically for the benefit of mankind. I am personally convinced that atomic energy will be of inestimable services to mankind in peacetime." De Gaulle, too, was eager to move the country forward and therefore asked Frédéric and Irene to craft a national plan for an atomic energy program.

Frédéric's goals were closely linked to the energy needs of France, which was by then forced to import nearly all of its fuel oil. In the next few years, Frédéric would go on to manage the construction of France's first atomic reactor, ZOE (Z for zero, as the power was small; O for oxide of uranium; E for *eau lourde*, or heavy water). The goal of the initiative, led by Frédéric, was to bring France to the point at which 80 percent of its electric energy would be generated within the country with the surplus exported for revenue.

At the top of a new Ministry of Atomic Energy was an Atomic Energy Commission, four of whose six members—Frédéric and Irene, Pierre Auger and Francis Perrin—were physicists and close friends. Frédéric, as high commissioner, was placed in charge of all scientific and technical work.

Irene also continued with her work as well as with her activism, at least as much as her uncertain health would allow. She spoke at a variety of events during the late 1940s, traveling to London, for example, to appear as a main speaker at an International Women's Day conference. More importantly, upon André Debierne's death in 1949, she succeeded to his post as director of the Curie Institute. She was finally at the helm of the institution where she had first worked in 1918.

Irene thrived there on her own with a team of sixty researchers to direct, and she was granted a full professorship at the Sorbonne in 1946. However, the tradition-bound Academy of Sciences was no more ready to accept a woman member in the mid-1940s than it had been during Marie's lifetime. But whereas Marie had felt deeply chastened upon her rejection, Irene took it in stride and remarked comically, "Well at least they are consistent in their thinking." Irene would try twice again, though, a not-so-subtle effort to remind people that sexual discrimination was still very much alive and well.

And now, with the war over, and work in France going well again, Frédéric was about to go to America for the first time. His political leanings would ultimately lead to his undoing.

CHAPTER 14

Rough Waters

In early 1947, Frédéric crossed the Atlantic for the first time, making a splash within days of his arrival by telling Americans that the French, too, knew how to make atomic bombs. Multiple media outlets quoted him as saying that his country most likely had the aptitude to do everything America could do. Writing in the magazine *United Nations World*, he also noted that France was moving toward the practical application of atomic energy to improve the living standards of mankind, and he urged other countries to do the same. Not only did he make clear that France had been a leader in the early stages of experiments on nuclear fission before the war broke out, but he also boasted that France's decision to turn its atomic program toward peace was "not based on any lack of knowledge about the atom bomb."

American scientists had for months warned that their country could not keep the nature of the atomic bomb a secret forever. And yet before Frédéric's visit most were of the opinion that—besides the United States—every other country's research was still in the experimental stages. After hearing from Frédéric, they were reportedly astounded at the strides France had made.

No doubt Frédéric's main intention was to return France to the status it had enjoyed before the war. Already, he had more than two hundred employees working for him at the French Atomic Energy Commission, a public body set up by President Charles de Gaulle in 1945. And he had been invited to come to America in 1947 as a delegate to the United Nations Atomic Energy Commission meeting at Lake Success, New York because of his high standing in scientific circles. Located on northwest Long Island, Lake Success was the temporary home of the United Nations from 1946 to 1951. The focus of discussions was to be the international control of nuclear energy. Again and again Frédéric talked of his country's program as being entirely directed toward peaceful efforts.

By this time, everyone knew that Frédéric was a devoted Communist. Much to his delight, the Communists had continued to grow in prestige in France after the war and, in the election of November 1946, received the most votes of any party. This political turn to the left, combined with Frédéric's frequent talk of how he intended to emulate the Russians—not to mention his long-time admiration for the way in which the Soviet Union generously funded its scientists—began to make the American government extremely uncomfortable. As a result, President Harry Truman, who supported a policy of strong resistance against Russian expansion, met with British Prime Minister Clement Atlee, and the decision was made to keep nuclear secrets from the French.

Interestingly, at a meeting at the United Nations, the American philanthropist and financier Bernard Baruch actively urged Frédéric to move to America. According to biographer Denis Brian, Baruch told him that "it is madness to try to do atomic energy in France. A pile—two piles—you'll never get them in the state your country is in. Your industry is not capable of providing what you need. You'd do better to stay in the United States." But in spite of all the promises of large staffs, modern facilities, and handsome salaries, Frédéric was not about to leave France.

On a personal level, Frédéric actually enjoyed New York City probably more than he had thought he would and was amazed by all the theaters, the bright lights, and the imaginatively decorated store windows. He wrote Irene that Manhattan reminded him of "Christmas time in Paris when I was a child . . . all those things in the windows that I would have liked."

At the invitation of Albert Einstein, Frédéric returned to the United States in September 1947 to attend, along with Irene, a commemoration of the two hundredth anniversary of the founding of Princeton University. Photos from this trip show a relaxed Irene and Einstein enjoying an informal conversation on the front steps of Einstein's home in Princeton, New Jersey.

These trips to America were something of a distraction for the couple, who had been grieving the death of Paul Langevin in December 1946 following a brief illness. The French government afforded him a state funeral, and his remains were transferred to the Pantheon in 1948 at the same time as those of Jean Perrin, another close Curie friend. At the funeral, an emotional Frédéric noted that Langevin was the one who "determined my destiny when he said he had a job for me with Madame Curie." Langevin had been Marie's lover, Frédéric's mentor, and Irene and Eve's close friend. His impact was considerable, and the relationship between the Langevins and the Curies was far from over. After meeting as university students, Langevin's grandson Michel Langevin and Irene and Frédéric's twenty-year-old daughter, Hélène, were married in 1948. Hélène later talked of how funny it was to watch her aunt, Eve, who at the time was in charge of a center-right newspaper, chatting at the wedding with Georges Cogniot, the director of a Communist paper. A wide range of personalities attended the wedding including Maurice Thorez, the longtime leader of the French Communist Party.

In the spring of 1948, Irene left on a third and final trip to the United States at the request of the American Joint Anti-Fascist Refugee Committee. (In 1946, the House Un-American Activities Committee had branded this committee as subversive.) The fundraising tour on behalf of Spanish citizens living in exile was a cause near and dear to her heart, but the tour got off to a rocky start when she was refused entry upon her arrival in New York City. Even though she'd obtained a valid two-week visa, her sponsors—as one might have expected—showed up on the US attorney general's list of subversive organizations. The adviser to the French United Nations delegation in New York had gone to the airport to meet Irene but wasn't even allowed to speak to her. The best he could do was wave through a glass partition and then stand by helplessly as immigration officials hauled her off to Ellis Island, where undesirable aliens were kept. Forever resilient, Irene kept her cool and spent the evening mending her stockings in a small room shared with three other women. Thanks to the intervention of the French Embassy, Irene was released the next day at noon. When a reporter asked about her reaction to the whole ordeal, she said she wasn't surprised. "I am here to aid the anti-fascists," she said. "This is not always as favorably considered as aiding the Nazis." She said that Americans look with much more favor on fascism than Communism, noting that "Americans think fascism has more respect for money."

A *Time* magazine article intimated that she actually had been treated better than she deserved since "all communists in a democracy are potential spies and traitors." The French government, of course, was outraged by her treatment. And members of the American Joint Anti-Fascist Refugee Committee—as well as Einstein—sent telegrams of protest to US government officials over Irene's detention. (Later Einstein launched a group called "The Committee of One Thousand" to seek the abolition of the House Un-American Activities Committee.)

But the rest of Irene's trip went off without a hitch. She was treated as scientific royalty by the general public everywhere she went, whether it was Boston, Chicago, or Philadelphia. In between stops, she even completed some work on a paper on the rays given off by radioactive ionium. Wherever she went, she urged more help for Spanish refugees. And she talked of the objectives of France, which was the first nation to establish a civilian atomic energy authority. She also lectured on the rights of women, as she had been doing for years in France, where she had been instrumental in helping women finally earn the right to vote in 1944, decades after women in the United States. And she had some fun on her American trip too. For example, witty writer Dorothy Parker hosted a warm and elegant reception for her French friend in New York City.

In August 1948, Irene attended a conference in Poland of "Intellectuals for Peace" and continued to push for more cooperation and contacts between intellectuals and artists on both sides of what had become known as the Iron Curtain. That same year, she also contributed to the creation of the first French atomic pile while serving as a member of the French

Atomic Energy Commission. (By this time, Frédéric was serving as High Commissioner.) It was a role she held until 1951.

But politics continued to get the pair into hot water.

One of the biggest brouhahas occurred in 1953 when Irene applied for membership in the American Chemical Society, one of the most important scientific associations in the world with 68,000 members. She filled out the application form as required, found current members to vouch for her character, and claimed she mostly wanted to become a member so that she could regularly receive the *Journal of the American Chemical Society*. The application never asked about her political affiliations as the society claimed it did not discriminate against applicants on the basis of race, color, gender, religion, or political beliefs. And yet her application was denied without anyone providing any reason for it.

Michael Heidelberger, one of the fathers of modern immunology in America, wrote a fascinating letter in 1953 protesting the decision to deny membership to Irene, saying it was done solely on the grounds that she was believed to be a Communist. Heidelberger, who had welcomed Irene during her visit to the United States in 1948, argued that the decision was entirely political and that it undermined the freedom of scientific inquiry throughout the United States.

Heidelberger said he agreed that Irene was "radical" but emphasized that she had never actually joined the Communist Party because "there was much that she did not like about it and she had too little time to work for it."

He noted, though, that he would agree with the admissions committee if they denied membership to Frédéric, not because he was an "avowed Communist," but because of his charge that the United States was guilty of using bacteriological warfare in Korea, "a subject upon which he could not possibly have had first-hand information in Paris."

Indeed, in the early 1950s, Frédéric drew criticism not only from American scientists but from other Nobel Prize–winning scientists all around the world when he published a letter in 1952 accusing not only US troops but also United Nations troops of using biological weapons in Korea. An issue of the *Bulletin of Atomic Scientists* published at least one article that was highly critical of Frédéric, charging that he had lent his name to the "scurrile accusation of biological warfare in Korea." That same article also noted that Frédéric had become one of the editors of a Communist journal called *La Pensée*, which celebrated the scientific qualities of Marxism.

"L'Affaire Curie," as the American Chemical Society episode was dubbed, dragged on into 1955. *Chemical Engineering News* printed an irate letter from Irene that described how she had always personally striven to overlook other scientists' political views and how warmly she had welcomed scientists of all nations—including those from the United States—into her laboratory. She said the members of the American Chemical Society should have been above such discriminatory behavior.

What she never mentioned, however, was that she and Frédéric had been urging Chinese scientists since 1949 to convince Chairman Mao Zedong's Communist regime to go forward with the construction of atomic and hydrogen bombs. Both Irene and Frédéric believed that, if it developed nuclear weapons, China could do good by breaking up the monopoly of nuclear powers like the United States while working behind the scenes to eliminate nuclear weapons.

After the Soviet Union successfully detonated an atomic bomb in 1950, President Truman publicly announced his decision to construct a hydrogen bomb that would be a thousand times more destructive than the atomic bombs dropped on Japan during World War II. Both the Americans and the British were staggered to learn that German-born Klaus Fuchs, a top-ranking scientist in the US nuclear program, had been a spy for the Soviet Union during and shortly after World War II. That meant, they believed, that the Soviets had as much access as Americans to information on how to build a hydrogen bomb. As a result, the race to complete the world's first "hell bomb" was on. And so was the Cold War.

As president of the World Peace Council established in 1949, Frédéric tried to organize a major peace conference in Sheffield, England, but the effort fell apart after the British government refused to allow any Soviet or Communist members into the country. The World Peace Council had basically been set up in response to a world that had—according to the group's founders—become increasingly divided between "a non-aggressive Soviet group and a war-minded imperialistic group, headed by the United States government." The council set up a separate World Committee of Partisans for Peace, chaired by Frédéric. He spoke often of how he truly admired the way in which the Soviets had always focused more on the social responsibilities of scientists.

But Frédéric's political leanings were leading him into trouble. In mid-March 1950, when he arrived in Stockholm to attend the inaugural meeting of Partisans for Peace, he was asked to leave the hotel where he had stayed as an honored guest only fifteen years before when he and Irene had been awarded their Nobel Prizes. The managers bluntly informed him that "Reds" were not welcome at their hotel.

If that wasn't bad enough, soon even the French government felt compelled to take action against him, especially with the US State Department increasingly wary of the scientist's politics. On April 29, 1950, a front-page story in the *New York Times* was headlined "French Remove Joliot-Curie as Chief for Atomic Energy." The story was true. Despite winning a Nobel Prize and having headed France's atomic program for so many years, Frédéric had been abruptly dismissed. The French government said that the decision was made not only because Frédéric was a Communist, but also because he was refusing to build atomic weapons. Indeed, when asked, Frédéric said that he would reply with an emphatic "no" if asked to work on anything other than a peaceful atomic program. Officials also said that Frédéric had been sabotaging arms shipments intended for French troops fighting in Indochina against the Viet Minh, a group battling for Vietnamese independence from French rule. Later, Frédéric's son, Pierre, was asked whether his father was surprised by his dismissal. He said he was, noting that his father believed he should be ruled only by his own conscience so long as he was performing his job at a satisfactory level. According to biographer Rosalynd Pflaum, Frédéric's son became an ardent socialist while his daughter became an active Communist. Although Frédéric's dismissal was protested by the Far Left, the general public seemed to approve of his removal.

But there were also some happy occasions for Frédéric during these years. He continued to lecture at the College of France, and both he and Irene celebrated the birth of their first grandchild, a girl named

Françoise, born to Michel and Hélène on May 21, 1950. In the coming years, Frédéric continued to write about his opposition to the possibility of American scientists building a hydrogen bomb and fought hard for peace as he continued to act as president of the World Peace Council and of the World Federation of Scientific Workers. (The World Peace Council is still a massive structure, with a presence in more than one hundred countries.) Both Irene and Frédéric publicly and frequently denounced the arms race and France's ongoing war in Vietnam—years before the United States became mired in the conflict.

But Frédéric had lost a lot. He no longer enjoyed the prestige and prominence of the years before. And even some of his closest friends turned away from him. His domination of French nuclear physics had come to a sudden close and France's progress lagged behind that being made in the United States and elsewhere. Some who knew him best complained that he was simply no longer a scientist by the late 1940s— he had become too much of a politician. Frédéric's health also worsened when he contracted hepatitis, which seemed to hang on longer than it should have. He confessed to a friend that he was certain he didn't have much longer to live and that he surely would die before his wife.

But that wasn't the case. By the early part of 1956, an increasingly frail Irene was being cared for in the Curie Hospital, where doctors delivered to her the good news that her tuberculosis was under control. Indeed, before her death, Meloney had sent Irene some streptomycin, an antibiotic drug that finally cured her tuberculosis. But the doctors then told Irene that, unfortunately, her frequent bouts of fever and dramatic weight loss were due to leukemia. Irene understood what this diagnosis meant but remained courageous despite the pain and even as she grew weaker. She told one close friend that she didn't fear death in the least because she'd enjoyed such a beautiful life.

Irene died of leukemia on March 17, 1956, at the age of fifty-eight. Her mind was sharp until the end but, despite all sorts of medications, she just couldn't hang on physically. Most agreed that her death had been caused by her work with X-rays during World War I and her later years of working with radioactive substances. What she would never know was how, in the years to come, leukemia would often be treated by injections of one or more of the thousands of radioactive isotopes deriving from her and Frédéric's discovery of artificial radioactivity. From all accounts, her death nearly destroyed Frédéric, who had lived and worked with this brilliant woman for thirty years. At the end, Frédéric had felt so poorly himself that he could only rally himself for very brief visits with his wife in the hospital. Eventually his health returned, though, and he was offered Irene's chair at the Sorbonne after her death—an offer he happily accepted. He also took over her role as head of the Curie Institute.

But Frédéric survived his wife by only two years. He succumbed to a liver disease caused, as his doctor put it, "from a cirrhosis produced by overexposure to polonium." Displaying his notoriously dark sense of humor, Frédéric had once called death from radioactive exposure "our occupational disease." Both of Irene and Frédéric's children were distraught over the loss of their parents, to whom they were quite close despite a string of lengthy separations over the years. Encouraged by Frédéric, a young Pierre had begun studying the ways energy from the sun's rays might be better utilized.

Often the two of them had enjoyed such lengthy and lively discussions at the dinner table that it had brought back memories for Irene of the many similar debates between Frédéric and Marie. Despite any previous disagreements, the French government decided to give both Irene and Frédéric state funerals. Although Irene probably would have preferred a small, quiet funeral similar to her mother's, in the end her coffin was displayed for twenty-four hours in the Sorbonne's main hall, where professors and students could walk by and pay their respects. The French government did honor Irene's pacifist views by forgoing the traditional military honors and guards. In various lengthy obituaries, Paris newspapers said that both deaths were a great loss to the country.

In an interview, Hélène said that she, as a scientist herself, only worked with her mother for about three months in total. "I never had the situation happen of me working that closely with her for a long period of time and she was always more of a mother to me than a friend," she said. "I was twenty-nine when my mother died. I had two children. I never discussed many things with her that I wish I had. You regret these things."

Hélène said her brother actually had more quality time with their mother. "I was at school and my mother had tuberculosis," she said. "My mother in those times spent a lot more time with my brother and they had more discussions together. My brother lived with my parents until their deaths. But of course I got married."

In 1955, one year before Irene died, Einstein had passed away in America of heart failure. Again, it was a death that saddened both Irene and Frédéric greatly. Einstein had always been a great supporter of the couple's political causes. For example, he was among those who had petitioned the US Supreme Court for a rehearing of the prison sentences— of between six and twelve months—imposed on John Howard Lawson and other members of the "Hollywood Ten" for contempt of Congress. Despite his criticisms of the Soviet Union, Einstein had always stood up for indicted Communists. It was his belief that Hitler's roundup of Communists had been the first sign that he was moving toward a concentration-camp society. Einstein had been close to Marie and had always remained close to Irene and Frédéric. Indeed, Frédéric was one of those mentioned in Einstein's 1939 letter to President Roosevelt as one of the leading scientists on the course to chain reactions.

When Einstein died in 1955 at Princeton Hospital, his body was cremated and his ashes were scattered, as he had wanted. But Thomas Harvey, the pathologist who performed the autopsy on Einstein before the cremation, decided there was a lot to be learned about why Einstein was so smart. During the procedure, he removed the brain to examine it, as he was supposed to, but then instead of placing the brain back into the skull, he put it in a jar of formaldehyde so that it could be studied later. He was fired from his position at Princeton because he refused to give up the brain. He kept it in mason jars and, every so often, sent a part off to one researcher or another for further study. Finally, in 1998, Harvey gave Einstein's brain back to the pathologist at Princeton Hospital.

Throughout all the years during which Irene and Frédéric were leaning more and more to the left, Eve was showing an even greater allegiance to America and to democracies around the world.

After the war, Eve went back to Paris, where she was hired as the copublisher of the daily center-right newspaper *Paris-Presse,* a job she held from 1945 to 1949. Next, she was appointed as the special advisor to the

Secretary General of NATO; she served in this role on the international staff until 1954.

One day in 1951, she met an American attorney and diplomat named Henry Labouisse Jr., who was at the time the chief of the Marshall Plan Mission to France, part of the effort to help Europe recover from the devastation of World War II. Ironically, while the United States believed a prosperous Europe was the most effective way to fight Communism, Irene and Frédéric had always felt that the Marshall Plan represented American interference in the internal affairs of other countries.

Labouisse left his position with the US government in 1954 to go to work for the United Nations. He was appointed director of the United Nations Relief and Works Agency for Palestine Refugees in June 1954 at the personal request of UN Secretary-General Dag Hammarskjöld. It was regarded as a dangerous and essentially thankless job, but Labouisse carried out his duties admirably. When he assumed the directorship, the program was responsible for taking care of 88,000 Arab refugees who had fled Palestine in 1948. He worked hard to improve their standard of living and organized a grant program so that refugees could make a down payment on a farm or a shop.

Eve fell madly in love with his dedication and commitment and the two married on November 19, 1954. Labouisse's first wife had died in 1945, and he had a daughter, Anne, from that marriage. Eve grew very close to her stepdaughter, who later gave her step-grandchildren and step-great-grandchildren.

Eve and Labouisse were quite similar in personality, sharing a passion for travel, adventure, and helping others—so they were a perfect match. Although he had been born in New Orleans, Labouisse was once described by his daughter as being "not of the mint-julep-and-magnolia school." He had been a hard worker in school, attending Princeton and then graduating from Harvard University Law School. During World War II, Labouisse decided to serve his country by accepting a position at the State Department, where he quickly rose through the ranks.

No matter what he did, though, his underlying message never shifted. He always felt that more should be done to help the poor. Eve shared his views and worked diligently at his side; she became a US citizen in 1958.

In 1960, as a consultant for the World Bank, Labouisse made various trips to Venezuela, devising a long-term economic program for the country that included a huge infusion of public investment. In a greatly publicized report, he advised using "the country's funds to make services available for all classes of people, including those in remote rural areas, rather than for the construction of more sophisticated hospitals."

A year later, he was appointed head of the International Cooperation Administration (ICA), a foreign aid agency established by the US government. Worried that it would be embarrassed if it sent inexperienced young diplomats into different cultures, government officials decided that Labouisse would be the perfect choice for the job. As part of the reorganization of foreign aid programs into the US Agency for International Development in 1961, the ICA was eliminated. The following year Labouisse was appointed US ambassador to Greece, a position he held until 1965. He and Eve grew quite attached to the country, even donating a parcel of land they owned there for a city park and constructing a kindergarten behind it with the last of the Marshall Plan funds.

But Labouisse's most important job would come next, when he was named executive director of the United Nations Children's Fund, or UNICEF, in June 1965. One of his first tasks was to accept the Nobel Peace Prize medal on behalf of the organization in Oslo on December 10, 1965. With Eve looking on, he memorably told the crowd that "The longer the world tolerates the slow attrition which poverty and ignorance now wage against 800 million children in developing countries, the more likely it becomes that our hope for lasting peace will be the ultimate casualty." Over the years he and Eve traveled to more than one hundred countries for UNICEF—the couple always covered Eve's costs—including such far-flung spots as south-central Tanzania, where they were given the honor of being made elders of the Wahehe tribe. Their last mission was to Cambodia, where UNICEF had initiated a widespread emergency relief program. This initiative entailed many visits to Vietnam. The couple also witnessed first-hand the horrific conditions of Cambodian refugee camps along the Thailand border, visits that affected both Labouisse and Eve greatly.

Although Labouisse retired in 1979, he and Eve retained a vital and affectionate interest in charity projects around the world, including the American Farm School in Thessalonika, Greece. This vocational training school for Greek rural youth had been founded by an American missionary in 1904. Labouisse served first as a trustee and then as chairman of the board from 1980 to 1985. He and Eve gave funding to the school and also donated a gymnasium and youth center. At the same time, Labouisse continued to consult on Thailand and Cambodia for UNICEF in the early 1980s.

From all accounts, Eve enjoyed a long and happy marriage to Labouisse, who died of cancer at Memorial Sloan-Kettering Hospital in 1987 at the age of eighty-three. Upon hearing of his death, a prominent UNICEF director said: "Harry Labouisse contributed so much towards making the world a better place for all of us, especially for our children." After his death, Eve divided her time between New York, Paris, and Greece although, for the most part, she was based in a New York City apartment.

In the end, Eve worked as hard as anyone for world peace over many years—exactly as her sister and brother-in-law always had. But their lives had played out very differently.

CHAPTER 15

The Legacy

"Life is not easy for any of us. But what of that? We must have perseverance and above all confidence in ourselves. We must believe that we are gifted for something, and that this thing, at whatever cost, must be attained."

—Marie Curie

It would be hard to imagine a family that has had the same kind of impact on the world as the Curies. Some might argue that it's an impact that was never fully appreciated until the latter part of the twentieth century.

Marie's reputation and influence were highlighted and reconsidered in 2011 when the world marked the one hundredth anniversary of her Nobel Prize in Chemistry. Calling Marie his personal hero, the actor Alan Alda added playwriting to his résumé with the opening of *Radiance: The Passion of Marie Curie*, a play that illuminated the challenges facing any female who sought a career in the sciences at a time when such a path was nearly unthinkable for women. On November 15, 2011, the world premiere of the opera *Madame Curie* by Elzbieta Sikora took place at the UNESCO House in Paris. Less than a month later, on December 10, 2011, Princess Madeleine of Sweden raised eyebrows when she announced that she would not attend the Nobel Prize award ceremony in Stockholm because she would prefer to attend the New York Academy of Sciences "Celebrating Women in Science" gala dinner to mark the centennial of Marie's second Nobel Prize. Also in 2011, universities across the country founded "Curie Clubs" to inspire their young female students to pursue careers in science. Science journals published commemorative issues dedicated to Marie and her work. Even Google paid tribute to Marie on November 7, 2011—on the 144th anniversary of her birth—by featuring a special Google Doodle in her honor.

But even before 2011, Marie's name had been pronounced again and again to evoke the wonders of research and discovery and exploration by science buffs around the world. When the American physicist Rosalyn Yalow shared the Nobel Prize in Physiology or Medicine in 1977 for research involving radioactive compounds, she said that Marie had been her inspiration. And many more the world has never heard of have been moved by Marie, not only by her story but by initiatives such as the Marie Curie Fellowships, administered by the European Union since 1996. These particular awards are so prestigious that recipients have often said that they are virtual career game-changers.

Meanwhile, there is a Marie Curie Middle School in Queens, New York, a Marie Curie School and a Marie Curie School for Medicine, Nursing, and Health Professions in Bronx, New York. And that's just in one city. There are also schools and streets named after Marie Curie in Kraków, Poland; Ottawa, Canada; Rheinbach, Germany; Dnepropetrovsk, Ukraine; and Bussy-Saint-Georges, France, among many other places.

Today you can buy Marie Curie postage stamps, gym bags, and pet bowls with the words "When I grow up I want to be a Marie Curie."

To find out more about the Curies, thousands of people flock to the Curie Institute Museum in Paris every year, which is located in the still very functioning Curie Institute, where scientific studies are still being conducted by doctoral and postdoctoral students.

Indeed, the Curie Institute today employs more than three thousand physicians and researchers treating more than thirteen thousand patients (new and ongoing). The institute's hospital is a reference point for the world when it comes to cancers and radiotherapy. And the Institute's research center is one of the most important in Europe and the largest in France devoted to cancer research.

When it comes to good works, the charity Marie Curie Cancer Care—which was launched in 1948 by doctors who wanted to preserve the Marie Curie name in the charitable medical field—is one of the biggest cancer charities in Britain. Employing more than 2,700 nurses, doctors, and other healthcare professionals, the charity provides care to around 25,000 terminally ill patients in the community and their hospices each year, along with support for their families. By 2011, the charity had opened nine hospices across England, Scotland, Wales, and Northern Ireland, and two centers for palliative care research. The charity also helps run the world-renowned Marie Curie Research Institute in England, which investigates the causes and treatments of cancer.

The Radium Institute founded by Marie in Warsaw, Poland, changed its name to the Maria Skłodowska-Curie Institute of Oncology after World War II. Today, it is a specialized health institute of the Polish Ministry of Health, the leading cancer research and treatment center in the country, with regional branches throughout Poland.

In France, too, where Marie was so vilified in the early part of the twentieth century, she has been honored again and again in recent years. In 1995, she became the first woman to win entry, on her own merits, to the Paris Panthéon, the nation's most sanctified final resting place. When President François Mitterrand ordered the remains of Marie—and also of Pierre—to be transferred to the Panthéon, it marked a symbolic victory for women's rights campaigners. During an emotional ceremony attended by ninety-one-year-old Eve, Mitterrand praised "the exemplary battle of one woman who decided to fight in a society dominated by men."

Ten years later, in April 2005, the main state television network in France invited viewers to vote for "The Greatest Frenchman of All Time." Viewers did not restrict themselves to men. Charles de Gaulle topped the list, with Louis Pasteur coming in second and Abbé Pierre third. Number four, and the first woman on the list, was Marie Curie.

Marie was certainly aware that she was a pioneer as a woman in science. Besides her Nobel Prizes, in 1906, Marie Curie became the first woman to be named not only lecturer and professor but also "head of the laboratory" at the Sorbonne University in Paris. But all of that was never important.

What motivated Marie was the sheer pleasure of the beauty of science and the enormous satisfaction derived from making the previously unknowable known. And she truly loved her daughters, tending to their intellectual and physical needs despite lengthy periods of separation. "I have frequently been questioned, especially by women, of how I could reconcile family life with a scientific career. Well, it has not been easy," Marie once said.

Of course, Marie also was the first Nobel Prize winner whose daughter also won a Nobel Prize.

Irene and Frédéric's discovery of artificial radioactivity has helped physicians with medical treatments since the 1930s. Just one example: clinics can manufacture radioactive versions of potassium and technetium, allowing a doctor to trace where the elements go in a patient. A sensitive detector then picks up the radioactivity outside the body and can locate and track the flow of blood and nutrients into certain organs.

Even those in fields such as geology benefit from artificially produced radiation as it helps them examine the minerals at local geological sites. After drilling a well and inserting a neutron source, that source will make the surrounding minerals briefly radioactive, and a detector subsequently measures the radioactivity. By studying the radioactive signals, geologists can figure out the density and makeup of the minerals.

There are even some buildings with exit signs lit with a radioactive version of hydrogen called tritium. In short, the uses of artificial radioactivity have been wide-ranging.

While Irene never received as many accolades as her mother, her work and her brilliance allow her to stand on her own among the greatest scientists of all time.

Irene always idolized her mother, saying that Marie never forced her to pursue science.

"That one must do some work seriously and must be independent and not merely amuse oneself in life—this our mother has told us always, but never that science was the only career worth following," she once said.

Even Eve, the one Curie not involved in the sciences, wound up having a huge impact as a biographer and a journalist—as well as a great humanitarian. Often called the First Lady of UNICEF, she travelled to dozens of the developing countries that received UNICEF assistance and was a tireless advocate for children and remained an honorary board member of the US Fund for UNICEF.

From all accounts, Eve had a wonderful sense of humor and took her family's fame in stride. "You are not mixing me up with my sister by any chance?" she said to a journalist who requested an interview in 1972. "You see, I am the only one of the family not to have won a Nobel Prize." She also sometimes joked that she had brought shame on her family. "There were five Nobel Prizes in my family," she said, laughing. "Two for my mother, one for my father, one for my sister and brother-in-law, and one for my husband. Only I was not successful."

She often referred to her simple background, noting that her mother was so modest Eve never even knew she was famous until their trip to America in 1921. But Eve made an impact all by herself, one that was recognized with several honorary doctorates. She received the Croix de Guerre for her work with Charles de Gaulle's Free French during World War II, and in 2005, she was awarded the rank of Officier de la Légion d'Honneur of the Republic of France—the country's highest decoration—by France's ambassador to the United Nations, Jean-Marc de la Sablière, for her work at UNICEF. At the time, Eve said she didn't believe she deserved such recognition, modestly admitting that she didn't "quite know how to behave." But in her younger days Eve was a spokeswoman for the women of France and she crisscrossed the United States, urging American involvement in World War II. "We discovered that peace at any price is no peace at all," she once famously remarked.

In December 2004, Eve marked her one-hundredth birthday by receiving UN Secretary General Kofi Annan in her Manhattan apartment. She also received warm congratulatory letters from both US president George W. Bush and French president Jacques Chirac.

Eve died in her sleep on October 22, 2007, at her home on the Upper East Side of Manhattan at the age of 102. She was survived by step-grandchildren and step-great-grandchildren. Her stepdaughter, Anne Peretz, told journalists that Eve felt enormous guilt that she alone among the women in her family had escaped a life of radiation and its consequences.

In a tribute following her death, UNICEF Executive Director Ann M. Veneman said that Eve's energy and commitment to the betterment of the needy should serve as an inspiration to the world.

"Mrs. Labouisse was a talented professional woman who used her many skills to promote peace and development," Veneman said. "While her husband headed UNICEF, she played a very active role in the organization, traveling with him to advocate for children and to provide support and encouragement to UNICEF staff in remote and difficult locations."

Today, with Eve gone, Hélène Langevin-Joliot is left to relay stories of her famous grandmother, grandfather, mother, father, and aunt. Speaking with her leads one to believe that being a member of the famous family is both a joy and a responsibility. Clearly she holds her grandmother in high esteem, noting that Marie's life "is an outstanding example of how science can be a human adventure." She has spent much of her adult life talking about her family, writing about her family, re-telling again and again stories of them and their discoveries. According to Hélène, her grandmother used to say that she dedicated much of her time to science because she loved research. And, for Hélène, that is the most important message. "She found a real joy working in the laboratory, researching the unknown, and finding her way to overcome many obstacles," Hélène said.

Born on September 17, 1927, Hélène has carried Marie's mantle in brilliant fashion. Not only is she director of research emeritus at the National Center for Scientific Research in Paris, but she also worked for years as a nuclear physicist at the Institute for Nuclear Physics.

She met and fell in love with Paul Langevin's grandson Michel Langevin when they both were students at the Paris Municipal School for Industrial Physics and Chemistry. Paris' scientific community is a closely knit one so it's not really all that unusual that the two met and married. In 1950, they had a daughter, Françoise (now Langevin-Mijangos), and the next year Hélène and Michel had a son, Yves. Françoise worked as an administrative director in the government's agricultural department. She is now retired. Yves became an astrophysicist. Michel Langevin, who was a nuclear physicist at the Curie Institute, died in 1986.

In 1954, Hélène helped to develop a scintillation spectrometer, which helps measure the energy distribution of particles given off in radioactive processes. Two years later she defended her thesis and by the end of the 1950s she had established herself as a leading scientist in France with her study of the polarization of electrons emitted through decay.

The year before her father's death in 1958, Hélène joined the faculty of the Institute of Nuclear Physics at the University of Paris at Orsay, where she became director of research in charge of a 580-person laboratory as well as a synchrocyclotron, a new-and-improved particle accelerator that was modernized under her direction. Like her parents and grandparents, Hélène has always been a pacifist and, also like all three generations of the Curie women, she lectures on the importance of genderless education and women's rights. Most importantly, she would like to see more women in science.

In an introduction to the book *Radiation and Modern Life,* Hélène wondered if Marie could ever have imagined how her scientific achievements would lead to such huge direct and indirect consequences. "What would she really think if she were alive today? We will never know. The only thing I can say with reasonable confidence is that she would be sad knowing that many people still fear radiation today, and that they are not aware of it enormous benefits."

As she admits, the age of the individual titans of radioactive science no longer exists. Teams of scientists from various countries now work together to achieve results from a single project. Hélène spent her career in particle physics and in the study of heavy nuclei. In recent years, computers have connected far-flung laboratories, meaning that Hélène's papers often have reflected the collective work and opinions of as many as twenty scientists working collaboratively worldwide.

In 1997, at the age of sixty-nine, Hélène made her own tour of America, seeking to encourage women to pursue careers in science during visits to schools and universities. She was particularly emphatic when it comes to the need to improve K–12 education. She has inspired her audiences by urging them to "work hard, adapt, and change the world for the better."

Today, she remains a sharp, energetic woman. And, like her mother and grandmother, she still believes that more time outdoors can soothe almost any ailment. After an interview, she said, "I am going off to Brittany after [this interview]. It is still so important to get fresh air."

Certainly, the world is changing. In 2009, Mariette DiChristina became the first woman editor in chief at *Scientific American*. And there have been a whole host of women scientists helping to change the world, ranging from Jane Goodall to Sally Ride to Shirley Ann Jackson. There is a sense that these women are standing on the shoulders of giants—the Curie women who have gone before.

It's an end of an era, really, with no Curie woman after Hélène to carry the family's torch. Her son, Yves, doesn't have any children while her daughter, Françoise, has a son, Antoine.

While Hélène doesn't seem to mind talking about her family, her brother Pierre, a biophysicist working on photosynthesis, has stayed more in the background.

When asked about his name, Hélène said he goes by Pierre Joliot. She noted how many people used to call her parents Joliot-Curie even though they signed their scientific papers "Irene Curie and Frederic Joliot." Hélène said her father asked her to sign Langevin-Joliot, instead of just Langevin, and that's what she has always done.

Pierre Joliot did travel to Calcutta, India, in late 2011 to deliver a lecture at a science institute inaugurated by his mother, Irene. Recalling his childhood, growing up in the company of Nobel laureate scientists, he told local reporters that "Everybody around me in the family was engrossed in science. After a while it became my passion. I enjoy the long hours of research in the laboratory like playing a game."

Displaying the closeness he had with his father, Frédéric, Pierre wrote in his 2001 book *The Passionate Research* that "It is moving for me to remember that shortly before his death my father spoke to an assembly of Nobel Prize–winning chemists at Lindau, [Germany,]when he expressed his nostalgia for a recent era when the basic nuclear physics experiments could be done on a wooden table in a small laboratory."

In India, Pierre talked about how the world needs support for long-term science projects.

"I think Pierre Curie wouldn't have been successful if he had worked now," he said. "There is no time for long-term research, everyone wants quick results."

And that brings this author back to how she became interested in this book in the first place. No doubt Marie's relationship with Meloney and with America in general is enormously fascinating—as is Eve's life history—and is uncharted territory. But, in the end, one can never forget that this story begins with a girl raised without a mother who eventually worked four years preparing a very tiny quantity of radium out of a messy ton of pitchblende uranium ore in order to prove there really was such an element. She had her foibles but she knew there was something there—somewhere—worth looking for. And if for no other reason, that is why "Manya" Marie Skłodowska Curie is one person worth learning more about.

SELECTED BIBLIOGRAPHY

I've listed several books and articles that were used as sources but not the hundreds of other works, papers, letters, theses, articles, and scientific journals that also were beneficial to me.

Biquard, Pierre. *Frédéric Joliot-Curie: The Man and His Theories.* Translated by Geoffrey Strachan. New York: Fawcett, 1966.

Brian, Denis. *The Curies: A Biography of the Most Controversial Family in Science.* Hoboken, N. J.: Wiley, 2005.

Chung, King-Thom. *Women Pioneers of Medical Research: Biographies of 25 Outstanding Scientists.* Jefferson, N.C.: McFarland, 2010.

Curie Archives, Paris.

Curie, Eve. *Journey among Warriors.* Garden City, N.Y.: Doubleday, 1943.

Curie, Eve. *Madame Curie: A Biography.* Translated by Vincent Sheean. New York: Pocket Books, 1965.

Curie, Marie. "The Discovery of Radium." Address by Madame Curie at Vassar College, May 14, 1921. Ellen S. Richards Monograph No. 2. Poughkeepsie, N.Y.: Vassar College, 1921.

Curie, Marie. "Nobel Lecture." Nobel Foundation, Stockholm, December 11, 1911.

Curie, Marie, and Irene Curie. *Correspondence, choix de lettres, 1905–1934.* Edited by Gilette Ziegler. Paris: Éditeurs français réunis, 1974.

Des Jardins, Julie. *The Madame Curie Complex: The Hidden History of Women in Science.* Women Writing Science. New York: Feminist Press, 2010.

Goldsmith, Barbara. *Obsessive Genius: The Inner World of Marie Curie.* New York: W. W. Norton, 2005.

Hurwic, Anna. *Pierre Curie.* Paris: Flammarion, 1995.

Kauffman, George B. "Marie Curie's Relations with the United States." *Chemistry International* 33, no. 1 (January/February 2011) .

McKown, Robin. *She Lived for Science: Irène Joliot-Curie.* New York: Julian Messner, 1961.*New York Times.* "Mayor Dedicates Marie Curie Ave." June 10, 1935 *New York Times.* "Mme. Curie Is Dead; Martyr to Science." July 5, 1934. *New York Times.* "Mrs. Mattingly, 82, Educator, Is Dead." January 7, 1934.

Nobelprize.org. Biographies of Marie and Irene Curie.

Ogilvie, Marilyn Bailey. *Marie Curie: A Biography.* Amherst, N.Y.: Prometheus Books, 2010.

Parker, Steve. *Marie Curie and Radium.* Science Discoveries. New York: Trophy, 1992.

Pasachoff, Naomi. *Marie Curie and the Science of Radioactivity.* Oxford Portraits in Science. New York: Oxford University Press, 1997.

Pflaum, Rosalynd. *Grand Obsession: Madame Curie and Her World.* New York: Doubleday, 1989.

Quinn, Susan. *Marie Curie: A Life.* New York: Simon & Schuster, 1995. Reid, Robert. *Marie Curie.* New York: New American Library, 1974 Roosevelt, Eleanor. "My Day." *Toledo Blade,* June 29, 1943. Rutherford, Ernest. *Collected Papers.* Published under the Scientific Direction of James Chadwick. London: Allen and Unwin, 1962.

Sharp, Evelyn. *Hertha Ayrton, 1854–1923: A Memoir.* London: Edward Arnold, 1926.

UNICEF press center. "UNICEF Mourns the Death of Eve Curie Labouisse." October 25, 2007.

Wilson, Steven Lloyd. "A Tale of Two Sisters: Eve and Irene Curie." *Pajiba Storytellers.* http://www.pajiba.com/pajiba_storytellers/a-tale-of-two-sisters-eve-and-irene-curie .php. April 6, 2011.

Wolke, Robert L. "Marie Curie's Doctoral Thesis: Prelude to a Nobel Prize." *Journal of Chemical Education* 65, no. 7 (July 1988): 561–573.

INDEX